AF598079

Methods in Molecular Biology™

Series Editor
John M. Walker
School of Life Sciences
University of Hertfordshire
Hatfield, Hertfordshire, AL10 9AB, UK

For further volumes:
http://www.springer.com/series/7651

Stem Cells and Aging

Methods and Protocols

Edited by

Kursad Turksen

Regenerative Medicine Program, Sprott Centre for Stem Cell Research, Ottawa Hospital Research Institute, Ottawa, ON, Canada

Editor
Kursad Turksen
Regenerative Medicine Program
Sprott Centre for Stem Cell Research
Ottawa Hospital Research Institute
Ottawa, ON, Canada

ISSN 1064-3745 ISSN 1940-6029 (electronic)
ISBN 978-1-62703-316-9 ISBN 978-1-62703-317-6 (eBook)
DOI 10.1007/978-1-62703-317-6
Springer New York Heidelberg Dordrecht London

Library of Congress Control Number: 2013931174

Printed on acid-free paper

Humana Press is a brand of Springer
Springer is part of Springer Science+Business Media (www.springer.com)

Preface

As our understanding of stem cells increases, it has become clear that changes in stem cells do occur during aging. Not only the changes in stem cell number are being reported but also the changes in their relationship to their microenvironment and their functionality as reflected in changes to their metabolome. With an aging population worldwide, understanding these age-related stem cell changes at a basic biology level and at the level of their impacts for regenerative medicine is of interest and importance. In this volume, I have brought together chapters with protocols critical for exploring the biology of stem cell aging. I am extremely grateful to the contributors for their generosity in sharing their expertise and time to describe details of their approaches.

I am, as always, very grateful to Dr. John Walker for his support of my interest in stem cell biology.

I acknowledge Patrick Matron for his commitment to this project and helping it to materialize.

A special *thank you* goes to David Casey for his outstanding efforts to help me complete the volume in a timely manner.

Ottawa, ON, Canada *Kursad Turksen*

Contents

Contributors

ILARIA BELLANTUONO • *Department of Human Metabolism, Mellanby Centre for Bone Research, University of Sheffield, Sheffield, UK*

ANTONIO PAOLO BELTRAMI • *Department of Medical and Biological Sciences, University of Udine, Udine, Italy*

CARLO ALBERTO BELTRAMI • *Department of Medical and Biological Sciences, University of Udine, Udine, Italy*

NATASCHA BERGAMIN • *Department of Medical and Biological Sciences, University of Udine, Udine, Italy*

REGINA BRUNAUER • *The Extracellular Matrix Research Group, Institute for Biomedical Aging Research, Austrian Academy of Sciences, Innsbruck, Austria*

ROSANNA CARDANI • *Dipartimento di Biologia molecolare e Biotecnologie, University of Milan, Milan, Italy; Centro per lo Studio delle Malattie Neuromuscolari—CMN, Milan, Italy*

DANIELA CESSELLI • *Department of Medical and Biological Sciences, University of Udine, Udine, Italy*

HUI CHENG • *State Key Laboratory of Experimental Hematology, Institute of Hematology and Blood Diseases Hospital, Chinese Academy of Medical Sciences, Tianjin, China; Peking Union Medical College, Tianjin, China*

TAO CHENG • *State Key Laboratory of Experimental Hematology, Institute of Hematology and Blood Diseases Hospital, Chinese Academy of Medical Sciences, Tianjin, China; Peking Union Medical College, Tianjin, China*

MARCO CHILOSI • *Department of Pathology, University of Verona, Verona, Italy*

MARIANA D. DABEVA • *Division of Gastroenterology and Liver Diseases, Department of Medicine, Marion Bessin Liver Research Center, Albert Einstein College of Medicine of Yeshiva University, Bronx, NY, USA*

FEDERICA D'AURIZIO • *Department of Medical and Biological Sciences, University of Udine, Udine, Italy*

RACHEL M. GERSTEIN • *Department of Microbiology and Physiological Systems, University of Massachusetts Medical School, Worcester, MA, USA*

MARZIA GIAGNACOVO • *Laboratorio di Biologia Cellulare e Neurobiologia, Dipartimento di Biologia Animale, University of Pavia, Pavia, Italy*

HATIM HEMEDA • *Stem Cell Biology and Cellular Engineering, Helmholtz-Institute for Biomedical Engineering, RWTH Aachen University Medical School, Aachen, Germany*

JOHNNY HUARD • *Departments of Microbiology & Molecular Genetics and Orthopaedic Surgery, School of Medicine, University of Pittsburgh, Pittsburgh, PA, USA; Stem Cell Research Center, University of Pittsburgh, Pittsburgh, PA, USA*

ANGELIKA JAMNIG • *The Extracellular Matrix Research Group, Institute for Biomedical Aging Research, Austrian Academy of Sciences, Innsbruck, Austria*

JI-WON JUNG • *Division of Intractable Diseases, Center for Biomedical Sciences, Korea National Institute of Health, Chungbuk, Republic of Korea*

Kyung-Sun Kang • *Department of Veterinary Public Health, BK21 Program for Veterinary Science, Adult Stem Cell Research Center, College of Veterinary Medicine, Seoul National University, Seoul, Republic of Korea*

Werner Kapferer • *The Extracellular Matrix Research Group, Institute for Biomedical Aging Research, Austrian Academy of Sciences, Innsbruck, Austria*

Sebastian Klepsch • *The Extracellular Matrix Research Group, Institute for Biomedical Aging Research, Austrian Academy of Sciences, Innsbruck, Austria*

Mauro Krampera • *Department of Pathology, University of Verona, Verona, Italy*

Mitra Lavasani • *Department of Orthopaedic Surgery, School of Medicine, University of Pittsburgh, Pittsburgh, PA, USA; Stem Cell Research Center, University of Pittsburgh, Pittsburgh, PA, USA*

Seunghee Lee • *Department of Veterinary Public Health, BK21 Program for Veterinary Science, Adult Stem Cell Research Center, College of Veterinary Medicine, Seoul National University, Seoul, Republic of Korea*

Günter Lepperdinger • *The Extracellular Matrix Research Group, Institute for Biomedical Aging Research, Austrian Academy of Sciences, Innsbruck, Austria*

Paulina H. Liang • *Department of Radiation Oncology, University of Pittsburgh School of Medicine, Pittsburgh, PA, USA*

Yi Liu • *Department of Physiology, Markey Cancer Center, University of Kentucky, Lexington, KY, USA*

Aiping Lu • *Department of Orthopaedic Surgery, School of Medicine, University of Pittsburgh, Pittsburgh, PA, USA; Stem Cell Research Center, University of Pittsburgh, Pittsburgh, PA, USA*

Manuela Malatesta • *Sezione di Anatomia e Istologia, Dipartimento di Scienze Neurologiche, Neuropsicologiche, Morfologiche e Motorie, University of Verona, Verona, Italy*

Patrizia Marcon • *Department of Medical and Biological Sciences, University of Udine, Udine, Italy*

Giovanni Meola • *Dipartimento di Neurologia, IRCCS Policlinico San Donato, University of Milan, Milan, Italy*

Sindhu T. Mohanty • *Department of Human Metabolism, Mellanby Centre for Bone Research, University of Sheffield, Sheffield, UK*

Laura J. Niedernhofer • *Department of Microbiology and Molecular Genetics, School of Medicine, University of Pittsburgh, Pittsburgh, PA, USA; University of Pittsburgh Cancer Institute, University of Pittsburgh, Pittsburgh, PA, USA*

Michael Oertel • *Division of Experimental Pathology, Department of Pathology, School of Medicine, University of Pittsburgh, Pittsburgh, PA, USA*

Carlo Pellicciari • *Laboratorio di Biologia Cellulare e Neurobiologia, Dipartimento di Biologia Animale, University of Pavia, Pavia, Italy*

Charusheila Ramkumar • *Department of Cell and Developmental Biology, University of Massachusetts Medical School, Worcester, MA, USA*

Kyle Rector • *Departments of Physiology, Markey Cancer Center, University of Kentucky, Lexington, KY, USA*

Stephan Reitinger • *The Extracellular Matrix Research Group, Institute for Biomedical Aging Research, Austrian Academy of Sciences, Innsbruck, Austria*

Mario Ricciardi • *Department of Pathology, University of Verona, Verona, Italy*

Paul D. Robbins • *Departments of Microbiology & Molecular Genetics and Orthopaedic Surgery, School of Medicine, University of Pittsburgh, Pittsburgh, PA, USA; University of Pittsburgh Cancer Institute, University of Pittsburgh, Pittsburgh, PA, USA*

ANNE SCHELLENBERG • *Stem Cell Biology and Cellular Engineering, Helmholtz-Institute for Biomedical Engineering, RWTH Aachen University Medical School, Aachen, Germany*

MAGDALENA SCHIMKE • *The Extracellular Matrix Research Group, Institute for Biomedical Aging Research, Austrian Academy of Sciences, Innsbruck, Austria*

SARVPREET SINGH • *The Extracellular Matrix Research Group, Institute for Biomedical Aging Research, Austrian Academy of Sciences, Innsbruck, Austria*

YU-ZHEN TAN • *Department of Anatomy, Histology, and Embryology, Shanghai Medical School of Fudan University, Shanghai, China*

SETH D. THOMPSON • *Stem Cell Research Center, University of Pittsburgh, Pittsburgh, PA, USA*

DANIELA TRIMMEL • *The Extracellular Matrix Research Group, Institute for Biomedical Aging Research, Austrian Academy of Sciences, Innsbruck, Austria*

GARY VAN ZANT • *Departments of Physiology, Markey Cancer Center, University of Kentucky, Lexington, KY, USA*

WOLFGANG WAGNER • *Stem Cell Biology and Cellular Engineering, Helmholtz-Institute for Biomedical Engineering, RWTH Aachen University Medical School, Aachen, Germany*

HAI-JIE WANG • *Department of Anatomy, Histology, and Embryology, Shanghai Medical School of Fudan University, Shanghai, China*

MLADEN I. YOVCHEV • *Division of Gastroenterology and Liver Diseases, Department of Medicine, Marion Bessin Liver Research Center, Albert Einstein College of Medicine of Yeshiva University, Bronx, NY, USA*

HONG ZHANG • *Department of Cell and Developmental Biology, University of Massachusetts Medical School, Worcester, MA, USA*

Chapter 1

Comprehensive Hematopoietic Stem Cell Isolation Methods

Kyle Rector, Yi Liu, and Gary Van Zant

Abstract

The use of flow cytometry has been critical in establishing methods to isolate and characterize hematopoietic stem cells (HSCs) and their progenitors. For more than 30 years, researchers have been uncovering novel markers that when used in combination significantly enhance the purification of HSCs from murine and human bone marrow. The complex interface between HSCs, the lymphohematopoietic system, and their niches, has made identification of HSC markers critical to understanding their biological nature, more so than other adult stem cell populations. Here we review the phenotypic markers and strategies used to purify HSCs, the appropriateness of using these markers for comparisons of HSC function at different stages of ontogeny, and their utility in defining the lineage bias in the HSC compartment.

Keywords Hematopoietic stem cells, Isolation methods, Phenotypic markers, Lineage bias

1 Introduction

Hematopoietic stem cells (HSCs) are defined by three cardinal characteristics that were originally postulated by Siminovitch and colleagues in early studies describing the radioprotective competence of these cells (1). First, these cells must be pluripotent and capable of giving rise to all mature myeloid and lymphoid lineages of the hematopoietic system. Second, although in steady state, HSCs are deeply quiescent, and when activated they must possess a high proliferative capacity. Third, and most important, in the presence of high proliferative and maturation cues, they must maintain the ability to self-regenerate to maintain their position at the apex of the hematopoietic hierarchy. Many of the seminal HSC studies and their results using whole bone marrow (BM) were confounded by the heterogeneity of the BM and unknown HSC phenotypes that precluded the absolute identification and isolation of cells responsible for long-term reconstitution of the hematopoietic system in irradiated animals. Thus, the rare and elusive nature of HSCs prompted the search for markers that would allow these cells to be separated and characterized.

Kursad Turksen (ed.), *Stem Cells and Aging: Methods and Protocols*, Methods in Molecular Biology, vol. 976,
DOI 10.1007/978-1-62703-317-6_1, © Springer Science+Business Media, LLC 2013

From a more contemporary point of view, many facets of HSCs have further expedited the discovery of novel markers needed to identify HSCs compared to other adult stem cell populations. Nearly all adult stem cells remain in a fixed anatomical position in close proximity to the progenitors and mature cell populations that they are responsible for replenishing, which facilitates easier histological examination and characterization (2). However, HSCs reside in a niche that is constantly being remodeled based on the developmental stage and energy demands of the organism, requiring flexibility to associate and dissociate with the stromal microenvironment. Moreover, HSCs can associate with different stromal niches (osteoblastic or vascular) presumably based on their activation or cell fate status (3). Unlike other stem cell populations, HSCs can migrate out of the BM altogether, move into the general circulation, and lodge into peripheral lymphoid tissues such as the spleen or lymph nodes. Thus, the dynamic and mobile behavior of HSCs has prohibited the discovery of a fixed locale and prompted the accelerated identification of unique phenotypic markers to purify and study HSCs at a population or single cell level.

2 Methods

2.1 Fluorescence Activated Cell Sorting Based Methodology to Isolate HSCs

The advent of flow cytometry has proven to be critical in the identification of mature hematopoietic cells due to the heterogeneous and undefined liquid nature of the hematopoietic organ. Using two different approaches, researchers attempting to identify HSCs have applied this technology to prospectively isolate HSCs to further characterize their function. One successful approach that identified cells with hematopoietic stem and progenitor cell (HSPC) characteristics was the use of fluorescently tagged monoclonal antibodies that bound specific cell surface proteins. Based on the differential binding of the antibodies to the surface of the cell (high, low, negative), researchers selectively isolated cells by fluorescence activated cell sorting (FACS) bearing a specific cell surface immunophenotype and functionally characterized them through in vivo transplant assays. The second approach that was being employed simultaneously was to select BM cells using supravital dyes that were nontoxic to cells. Using a similar approach to immunophenotypic methods, cells were isolated based on the same high, low, and negative staining criteria and functionally characterized. Although recent advances have been made in the purification of human HSCs (4), seminal findings using FACS-based strategies critical in advancing the isolation of HSCs are reviewed here with a restricted emphasis on murine cells.

2.2 Immunophenotypic Isolation of HSCs

The first successful attempt to immunophenotypically purify HSPCs was published in 1988 by Irving Weissman's laboratory which involved a first step that depleted mature specialized cells (B-cells,

T-cells, granulocytes and macrophages) (5). This approach has been adapted by many labs using a cocktail of biotinylated antibodies specific to receptors expressed on mature lineages of hematopoietic cells (CD5/Ly-1=Lymphocytes, CD45R/B220=B-cells, CD11b/Mac-1=Macrophages, Ly-6G/Gr-1=Granulocytes, Ter-119/Ly-76=Erythrocytes, CD8a/Ly-2=T-cells, CD4/Ly-4=T-cells). Labeling cells with streptavidin conjugated with a single fluorochrome provides a simple method to selectively "lineage deplete" (Lin-/lo expression) all cells that express markers for mature hematopoietic cells, leaving multiple channels open for other markers of HSCs. The first steps in this strategy are depicted in Fig. 1a, b. Figure 1a depicts a population with low side scatter and low forward scatter, indicating that the HSPC population is relatively small and has a cytoplasm of low complexity or granularity. Fig. 1b depicts gating of the Lin-/lo population of cells. In their purification strategy, they added an additional marker, Sca-1, which further reduced the population to ~0.5% of the BM. They successfully showed that 30 Lin^-Sca1^+ (LS) cells were capable of saving 50% of the lethally irradiated mice that were transplanted. This work demonstrated that cell surface markers could phenotypically define the relatively rare HSPC population, either through positive or negative selection and thus established that HSCs are not phenotypically null cells that receive cues from the external environment to differentiate into specific hematopoietic lineages. The two markers used in this study remain staples of almost every HSC purification strategy used today.

This FACS-based purification strategy was rapidly expanded over the course of the next few years. In 1991, c-Kit, the receptor for stem cell factor, was identified as a positive marker for the purification of hematopoietic progenitor cells (HPCs) (6). c-Kit used in combination with Lin^-Sca^+ (LSK) further enriched the HSC population, which represented approximately 0.05% of all the cells in the BM as depicted in Fig. 1c (7). Building on previous work that showed some primitive hematopoietic progenitor populations express varying levels of the Thy1.2 antigen (8–11), $Thy1.1^{lo}$ was then used in combination with LS or LSK strategies to demonstrate that this phenotypic population contained all of the long-term repopulating HSCs (12, 13). However, the $Thy1.1^{lo}$ phenotype is exclusive to HSCs of mouse strains that carry this allele. Because mouse strains carrying the Thy1.2 allele express variable levels of the allele, Thy1.2 is not a reliable marker for HSC purification (14).

As the underpinnings of the HSC immunophenotype were being established, considerable heterogeneity still existed in the engraftment kinetics and multilineage potential using the aforementioned purification strategies. This prompted Morrison and Weissman to further fractionate the LS $Thy1.1^{lo}$ population into $Lin^-Mac\text{-}1^-CD4^-$; $Lin^-Mac\text{-}1^{lo}CD4^-$; and $Lin^-Mac\text{-}1^{lo}CD4^{lo}$ to determine if the heterogeneous reconstitution capacity could be delineated based on excluding cells with low levels of mature lineage markers (15). These studies successfully demonstrated that

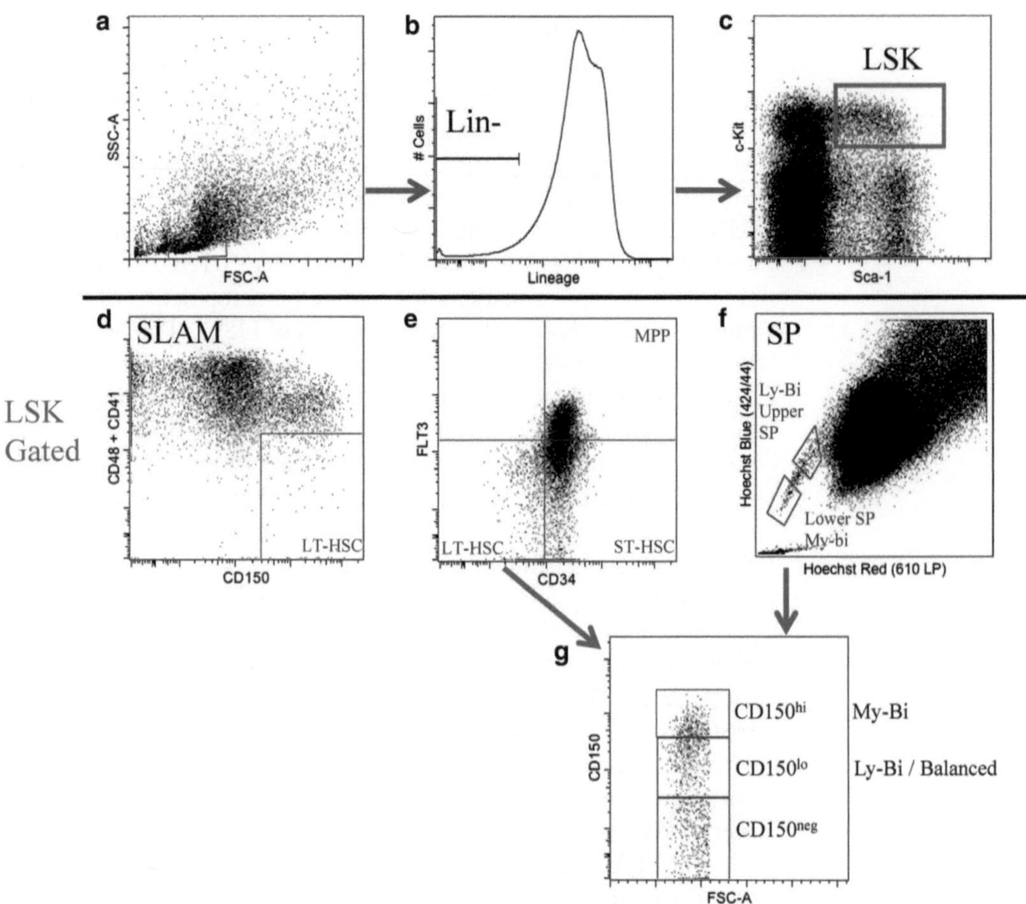

Fig. 1 FACS plots depicting the common methodologies and gating strategies for isolation of HSCs. The *top* three plots depict the LSK gating strategy common to most isolation methods. The plots are as follows from *left* to *right*: (**a**) Gating on HSPC (a.k.a. lymphocyte) region of forward scatter vs. side scatter plot, (**b**) selection of lineage negative/low population, (**c**) gating of Lin$^-$cKit$^+$Sca1$^+$ population that contains all of the long-term (LT) and short-term HSCs. The LSK population can be further purified via three different methods to yield a highly enriched population of LT-HSCs. (**d**) SLAM gating strategy with CD150 on the *x*-axis and both CD41 and CD48 staining on the *y*-axis as the HSCs are negative for both markers. (**e**) CD34 staining on the *x*-axis and Flt3 on the *y*-axis. Quadrant gating allows three distinct populations of multipotent cells to be separated. The CD34$^-$Flt3$^-$ population can be further fractionated by CD150 positivity (**g**) to select for myeloid and lymphoid biased HSCs, (**f**) Side population is gated on the lymphoid biased upper SP and myeloid biased lower SP. Each of these populations can be further subdivided (**g**) to further enrich for HSCs with a specific lineage bias

the long-term repopulating cells were all Lin$^-$Mac-1$^-$CD4$^-$, while the other two populations were short term reconstituting. While this did not establish a clear immnuophenotype of the long-term (LT)-HSC, it did provide supporting evidence that a clear, predetermined hierarchy existed in the HSPC population.

The single most important proof that HSCs with long-term engraftment were isolatable by immunophenotype came from studies by Nakauchi and colleagues (16). They utilized a novel marker, CD34 to subdivide the LSK population into two populations: 90% of which were CD34^{+} and the remaining 10% were CD34^{-}. Selecting cells bearing the CD34^{-} phenotype in the LSK population represents approximately 0.005% of BM cells. We note that these percentages reflect those seen in young 8–12 week old C57BL/6 mice and this increases considerably with age. *Also, CD34 staining is atypical as it requires a minimum 90-min incubation at a relatively higher concentration than most other antibodies.* Selectively isolating LSK CD34^{-} cells allowed this group to define a profoundly purified HSC population that when transplanted as single cells into lethally irradiated mice provided long-term reconstitution in ~20% of all mice transplanted. The CD34^{+} cells provided only short-term reconstitution, which was counterintuitive because all HSC activity in humans is contained within this population. Therefore, this study subdivided the murine LSK population into LT-HSC that are LSK CD34^{-} and multipotent progenitors that are LSK CD34^{+}. Moreover, these findings made a significant contribution to the ability to purify a more homogenous population with long-term repopulating potential.

FLK2 expression was later shown to be associated with a loss of primitiveness in the LSK population and its restricted multipotent progenitors (17, 18). Combining this marker with CD34 allowed the LSK population to be parsed into three sub-populations with uniquely different engraftment potential based on their immunophenotype, depicted in Fig. 1e (19). This is currently a common and highly efficient strategy used to isolate and characterize multipotent stem and progenitor cells in various mouse models.

LSK CD34^{-}FLK2^{-}: LT-HSC
LSK CD34^{-}FLK2^{+}: Short-term HSC
LSK CD34^{+}FLK2^{+}: Multipotent Progenitors

While each population is progressively more differentiated and capable of at least transient multilineage reconstitution, the CD34^{+}FLK2^{+} population is seemingly more primed for higher lymphoid potential, whereas the LSK CD34^{-}FLK2^{+} population is primed for myeloid cell output (20, 21).

In 2005 Morrison's group used a more contemporary approach to identify markers that would distinguish between true LT-HSCs and multipotent progenitors (22). They used gene expression arrays to compare differentially expressed genes among long-term and transiently reconstituting HSCs and identified a unique signature in the expression of genes from the signaling lymphocyte activation molecule (SLAM) family of cell surface markers. Based on the differential expression of CD150, CD244, CD48, they also characterized three populations within the HSPC hierarchy that they dubbed the SLAM code.

CD150^{+} CD244^{-} CD48^{-}: LT-HSCs
CD150^{-} CD244^{+} CD48^{-}: Multipotent Progenitors
CD150^{-} CD244^{+} CD48^{+}: Highly Restricted Progenitors

This simple three-marker approach ($CD150^+$ $CD244^-$ $CD48^-$) enabled the isolation of single long-term engrafting HSCs with the same engraftment potential as when using a complex LSK $CD150^+CD48^-$ sorting scheme. Figure 1d illustrates how the LSK gating scheme can be combined with SLAM markers to sort the LT-HSC population of cells.

CD150 unequivocally identifies HSCs and has been used extensively with other strategies to further purify HSCs to homogony, which typically includes any combination of LSK and CD34, LSK and FLK2, or more recently LSK, CD34, and FLK2. In addition, the endothelial protein C receptor (EPCR) (23) has shown marked purification of HSCs when used in combination with the SLAM cocktail of antibodies (E-SLAM) or any of the aforementioned strategies (24).

2.3 Dye Exclusion Methods for Purification of HSCs

The exclusion of fluorescent dyes is another approach that has proven advantageous to identify HSCs from the complex mixture of BM cells. The advent of fluorescent dyes was extremely useful as it allowed HSPCs to be selected based on the activity (multi-drug resistance (MDR) pumps) of the cell rather than expression of cell surface proteins. This is perfectly logical as most HSCs remain in a dormant or deeply quiescent state compared to actively cycling multipotent progenitors (25, 26). Moreover, the dye would provide a simple method for interspecies comparison of HSC functional attributes. Several studies (27–32) successfully used the nontoxic mitochondrial specific dye Rhodamine123 (Rho123) to purify and characterize HSPCs with long-term repopulating ability. $Rho123^{hi}$ gave rise to early colony forming unit-spleen (CFU-S) colonies and displayed minimal long-term repopulating ability. However, cells that poorly stained with Rho123 showed a robust ability to provide radioprotection to the recipient mouse and gave rise to no early CFU-S Day 8 colonies and few late Day 12 colonies. Exclusion Rho123 staining provided a simple approach to significantly enrich HSCs with a high repopulating efficacy. It is not completely clear how Rho123 differentially stains primitive HSCs, but it likely results from a combination of highly active MDR pumps present in HSCs as well as differences in mitochondrial membrane potential that prevents accumulation of Rho123 in HSCs (33).

A second approach commonly employed to isolate HSCs was developed in 1996 by Goodell and colleagues using the vital fluorescent dye Hoechst 33342. When Hoechst 33342 is excited by UV laser, it simultaneously emits light at two wavelengths approximately 450 nm (Hoechst Blue) and 675 nm (Hoechst Red). Analysis of Hoechst Blue vs. Hoechst Red on a two-dimensional plot revealed an interesting continuum of cells that selectively excluded the Hoechst dye, called side population (SP) depicted in Fig. 1f (34). Isolation and transplantation of the SP markedly enriched HSCs about 1,000-fold in the absence of any other markers. Similar to Rho123 staining, efflux of Hoechst 33342 from HSCs in the SP depends on the expression of MDR pumps as inhibition of MDR activity by the specific inhibitor Verapamil dissipates the SP phenotype. While SP and Rho123 labeling is extremely effective in enriching HSC activity, it is more common to couple these strategies with complex immunophenotypic methods that more stringently select for HSC potential, e.g., SP-SLAM, SP-LSK (Fig. 1f), or any combination of CD150, FLK2, CD34, and EPCR.

While both approaches involving dye exclusion are simplistic in nature and significantly enrich HSCs these procedures rely on the biological activity of the cells and their ability to selectively exclude the dyes. *These processes are time, temperature, and concentration dependent; therefore, procedural optimization and consistency are critically important for inter-experiment or inter-lab comparisons* (35).

2.4 Cell Surface Marker Expression with Regard to Development, Hematopoietic Stress, and Age

HSCs recovered during murine fetal development from the fetal liver (FL) have a much greater potential for reconstituting the hematopoietic system compared to HSCs derived from the adult (36, 37). This has great implications for our understanding the ontogeny of HSCs that will prove beneficial clinically as umbilical cord blood, sought after birth in humans, is currently a viable and rich source of HSCs. Characterizing the absolute engraftment characteristics of FL HSCs compared to adult HSCs critically depends on the ability to identify and isolate these cells to homogeny, as the cellular composition in the FL is much different from BM in an adult mouse.

The first strategy used to immunophenotypically characterize the FL population identified AA4.1 as a cell surface marker positively identifying all primitive FL HSCs, which constituted 0.1% of all FL cells with long-term repopulating potential (38). Combining this marker with conventional strategies to purify a population with $Lin^{-}AA4.1^{+}$ $Sca\text{-}1^{+}$ immunophenotype further refined this strategy (39). The AA4.1 antigen was absent from the long-term $Lin^{-}c\text{-}Kit^{+}Rho123^{lo}$ adult BM cells (40). Thus, AA4.1 expression was specific to FL HSCs, which demonstrated the first fundamental immunophentypic differences between adult and FL HSCs. A different isolation strategy used by Morrison and coworkers demonstrated that $Lin^{-}Thy\text{-}1^{lo}Sca\text{-}1^{+}Mac\text{-}1^{+}CD4^{-}$ FL cells are HSCs that have long-term multi-lineage reconstitution capacity (36). This method contrasts to their previous studies (15) that showed the most pure population of HSCs in adult BM was negative for the mature lineage marker Mac-1, demonstrating a second clear difference between adult and FL HSCs in their expression of cell surface markers. CD34 negativity of adult LSK cells has been a favored approach to separate LT-HSCs; however, fetal HSCs are uniformly positive for this antigen through all stages of development (41, 42).

HSCs remain CD34^{+} through fetal development, neonatal development, until approximately 7–10 weeks after birth (43, 44), therefore, caution must be used when determining an isolation strategy of HSCs during the adolescent stages of mouse development. The SP and Rho123 efflux phenotype also appear to display a similar pattern into adulthood (45). These dye efflux strategies are quite effective at isolating adult HSC populations; however, this seems to be a poor strategy to isolate HSCs through the early stages of development as the SP and Rho123 low fractions yield a fraction of cells with heterogeneous long-term reconstitution kinetics. The phenotypic switch of HSCs into the SP and Rho123 low compartments occurs at approximately 4 weeks after birth, corresponding with fully developed adult immunity.

These findings highlight the importance of understanding changes during developmental stages and activation status of HSCs. However, they also reflect a tremendous need to determine a stable panel of markers for the direct comparison of highly enriched HSC populations independent of cell status. The SLAM family of cell surface markers has provided the most uniform strategy to identify HSCs to directly assess differences in engraftment potential between cells at different stages of ontogeny (46). Isolating FL HSCs using their previous method to isolate adult HSCs, they showed that these markers are stable between fetal development and adulthood. However, only 18% of cells bearing this phenotype were LT-HSCs. This contrasts to the 45% observed to be LT-HSCs in adult mice using the same panel of SLAM markers (22). Selecting for the markers Lin^{-}, Sca-1^{+}, and Mac-1^{+} to the SLAM panel will further enrich a population to contain approximately 37% long-term HSCs. These results suggest that, to more accurately compare engraftment potential, developmentally conserved markers need to be screened to elucidate previous FL studies, which document their high reconstitution potential compared to adult HSCs. Identifying markers that are differentially expressed between FL and adult HSCs is interesting as it alludes to different activation patterns in the HSCs. These differences can be attenuated in adult HSCs by administering 5-fluoruracil (5-FU) to kill actively cycling progenitor cells and promote massive proliferation and expansion in the HSC compartment. Reversible expression has been described for CD34, Mac-1, and CD38 receptors, in addition to the SP and Rho123^{lo} phenotypes. Upon activation after administration with 5-FU, adult HSCs with the highest reconstituting potential are CD34^{+}, Mac-1^{lo}, and CD38^{+}, thus mimicking a FL HSC phenotype (47–49). CD34 expression on HSCs was examined after hematopoietic reconstitution in recipient mice reaches steady state, and the fraction of HSCs with long-term reconstitution reverts back to a CD34^{-} phenotype, as demonstrated when CD34^{+} and CD34^{-} fractions were isolated from the primary recipients and transplanted into secondary hosts (48). The same effect is observed when characterizing the SP and Rho123^{lo} populations after 5-FU treatment (45), which yields a heterogeneous SP and Rho123^{lo} fraction with varying reconstitution capacities. c-Kit expression is also reversible on HSCs from 5-FU treated mice, with the highest reconstituting potential in the c-Kit$^{-/lo}$ fractions (47), which significantly contrasts with both steady state adult HSCs and FL HSCs that show high levels of c-Kit expression. The coordinated developmental HSC phenotypes through adolescence and reversion after 5-FU treatment has been suggested as a potential link to an epigenetic master switch that exists to regulate the activity of HSCs during periods when the demands of the hematopoietic system is greatly increased (45). However, downregulation of the c-Kit receptor may indicate that the association with adult HSC activation state and FL phenotype is coincidental.

As an organism ages, it is presumed that the quality of individual HSCs change as they are continuously encountering cytotoxic insults throughout the lifespan. Using immunopheotypic or dye efflux methods, expansion of the defined HSC has been observed, however this expansion is without a concomitant increase in HSC frequency as measured by long-term repopulation of isolated cells (50–52). Morrison and colleagues (51) showed that approximately 20% of LSKT^{lo} cells from young donors give long-term multi-lineage reconstitution in recipient animals, however using the same panel of markers, approximately 1.3% of LSKT aged donor cells successfully engraft long-term It may be assumed that detrimental changes have taken place to alter the ability of the HSC to engraft, however these differences have been reconciled by selecting stable markers in the purification of HSCs with more absolute function. The SLAM markers, when used in combination with a LSKT marker panel, yield a population of old donor cells that reconstituted at a 14% efficiency (53), compared to 50% for young donors. This suggests that the discrepancy in the reconstitution potential in older donors is not as great when using high stringency methods to isolate true HSC function.

2.5 Parsing the Heterogeneity and Lineage Bias with SP and CD150 Expression

Although stringent methods of FACS-based purification have isolated a homogenous long-term engrafting HSC population, a significant amount of heterogeneity in the mature progeny output exists at the population level (54–56). The Mueller-Sieberg laboratory has been instrumental in defining the existence of clonal lineage bias in the HSC compartment (57–59). They demonstrated that single HSCs, when transplanted, preferentially give rise to mature cells skewed toward myeloid (termed My-bi) or lymphoid (termed Ly-bi) HSCs lineages. Moreover, the epigenetic memory of lineage preference is retained in all clones generated from the original HSC (60). Distinct phenotypes have not been uncovered that can easily distinguish between the My-bi and Ly-bi HSCs, but studies from the Goodell, Nakauchi, and Rossi labs (61–63) have provided convincing evidence that the heterogeneity in the HSC pool can be parsed out either by the level of dye efflux and/or CD150 expression. These studies used slightly different strategies to identify lineage biased HSCs, but all reached the conclusion that heterogeneity in the HSC population can be prospectively determined by phenotypic measures. Two different labs used similar approaches to identify HSCs by using either the LSK CD34$^-$ (63) or LSK CD34$^-$FLK2$^-$ (61) phenotype and then separating these cells according to their CD150 expression of CD150$^-$, CD150^{lo}, or CD150^{hi}, as indicated in Fig. 1.1e–g. When transplanted, significant bias was observed between the populations, with CD150^{hi} preferentially giving rise to myeloid lineages and CD150^{lo} to lymphoid lineages. Notably, in both studies, a balanced HSC population existed in the low/intermediate staining CD150 population. Aged mice showed a stark expansion in the CD150^{hi} population, which seemed to possess the most robust self-renewal capacity in serial transplantation studies, likely contributing to the myeloid lineage skewing in the peripheral blood of aged animals.

Strengthening the notion that HSC subtypes could be prospectively isolated, a different approach was used to separate the LSK side population into lower SP and upper SP based on the differential ability of cells in the SP to efflux Hoechst 33342 dye (62). This method had previously been used to show that lower SP cells have a much higher propensity for long-term engraftment (64, 65). Transplantation studies of these two populations successfully segregated My-bi cells into the lower SP fraction and Ly-bi cells into the upper SP fraction, as indicated in Fig. 1.1f–g. CD150 was also used to further enrich HSCs with bias in the upper and lower fractions of the SP. CD150 positivity associated with higher My-bi potential in both populations, albeit to a lesser degree in the upper SP. In contrast, CD150 negativity was associated with a moderate increased preference for Ly-bi production in both fractions of the SP. CD150 selection of cells in the SP was based strictly on presence or absence of the marker, unlike Beerman et al. (61) and Morita et al. (63) strategies where CD150 selection was based on the level of expression on the cell surface. In aged mice, expansion of the lower SP followed the same trend as expansion of the CD150 population, thereby supporting the validity of the two approaches in their characterization of the change in the composition of the HSC compartment with age.

The current methods of HSC purification are intriguing and it is becoming increasingly clear the power of isolating true LT-HSC increases with increasing complexity of the number of phenotypic markers used. Whether or not a CD150$^-$ HSC population exists has been hotly debated (66, 67). The Morrison group has suggested that all of the long-term engrafting cells reside within the CD150$^+$ population. Purifying HSCs with more stringent methods (LSK CD34$^-$ or FLK2$^-$) and comparing reconstitution potential in the CD150$^+$ and CD150$^-$ sub-fractions would be more appropriate, and recent transplantation experiments using these parameters have argued that a subpopulation of CD150$^-$ cells are multi-lineage reconstituting HSCs. We are not disputing the robustness of CD150 as a marker to identify the LT-HSC population; however, when using stringent purification methods such as the LSK SP, LSK CD34$^-$FLT3$^-$, or E-SLAM strategies to isolate LT-HSCs, it is clear that the subpopulation of CD150$^-$ cells are capable of some self-renewal ability (24, 62, 63). This suggests a highly complex continuum of marker expression within the LT-HSC population that requires further investigation. Beneveniste et al. showed that CD49b positivity associates with short-term repopulation kinetics and its expression mostly segregates the CD150$^-$ population from CD150$^+$ (68). Although they did not select CD150 expression in their transplantation studies, inclusion of CD49b into a sorting scheme could ultimately resolve the CD150$^-$ population into HSCs with intermediate and short-term reconstitution potential. While it remains controversial whether CD150$^-$ cells are true LT-HSCs, it can be said conclusively that CD150^{hi} population are at the apex of the hematopoietic hierarchy, and by virtue of its epigenetic pleiotropy gives rise to both the CD150^{lo} and CD150$^-$ populations, along with a predetermined preference towards myeloid effector lineage output.

2.6 Conclusion

Identifying HSC markers has not been a trivial process. However, a massive effort has identified a multitude of cell surface markers, making it possible to isolate the long sought after primitive LT-HSC. Combining various isolation strategies has allowed researchers to identify HSCs at different stages of ontogeny and myeloablative conditions. However, with the exception of the SLAM family of markers, most cell surface proteins display dynamic expression patterns that hinder the ability to isolate and comparatively characterize HSCs at different stages of development. Thus, when isolating HSCs, the developmental status of the animal must be considered to ensure that markers will positively select HSCs under those circumstances.

We are not advocating for a "best strategy" to functionally characterize HSC behavior, but it is important to use caution when interpreting one's own results or performing inter-lab comparisons of transplant data. This is especially true when comparing the biological behavior of HSCs from old and young mice, as the functional heterogeneity of the HSC compartment appears to be lost with age and skewed towards myeloid biased HSCs. Interestingly, the intrinsic robust self-renewal potential and expansion of the My-bi HSC compartment may be related to its vigorous dye-exclusion properties, which allows it to withstand the effects of the aging. The existence of the My-bi HSCs in the lower SP suggests it has a much higher capacity to extrude potentially cytotoxic or genotoxic agents from the cell more efficiently than Ly-bi HSCs via MDR pumps. Does a higher activity of the MDR transporter give a selection advantage to My-bi HSCs through the natural aging process? This profound ability to escape the selective pressures of the aging process could permit a more faithful self-renewal program that promotes a higher capacity to compete for existing niches and expand with age. In contrast, their Ly-bi counterpart could potentially sustain more DNA and cellular damage with age resulting in a diminished capacity for self-renewal and ultimately loss via attrition or cellular senescence mechanisms.

Additionally, it seems prudent that the definition of a HSC at the clonal level, and not population level, be reconsidered. After isolation and transplantation of prospective Ly-bi and My-bi HSCs, some clones produce virtually no detectable levels of the other lineage suggesting that the HSCs can exclusively give rise to either lymphoid or myeloid lineages. While these cells are pluripotent and can give rise to multiple lineages, Ly-bi (B-cell, T-cell, NK cell) and My-bi (granulocytes, monocytes, platelets, RBCs), it is not an absolute requirement to give rise to all hematopoietic lineages to be considered an HSC.

The most recent progress in stem cell purification methods has laid the groundwork to assess the complexity of the hematopoietic hierarchy at the LT-HSC level. Furthermore, these advances have provided an important step in identifying age related changes in the HSC compartment that correlates extremely well with expansion of the mature myeloid cells throughout the organismal aging process. Inclusion of these new techniques can be used to refine current protocols to permit further parsing of the genetic/epigenetic components that dictate fates of biased HSCs at the cellular level.

Acknowledgments

We would like to thank Paula Thomason for her expert editorial assistance in the preparation of the manuscript and Ying Liang for her critical review.

References

1. Siminovitch L, McCulloch EA, Till JE (1963) The distribution of colony-forming cells among spleen colonies. J Cell Physiol 62:327–336
2. Snippert HJ, Clevers H (2011) Tracking adult stem cells. EMBO Rep 12:113–122
3. Yin T, Li L (2006) The stem cell niches in bone. J Clin Invest 116:1195–1201
4. Notta F, Doulatov S, Laurenti E, Poeppl A, Jurisica I et al (2011) Isolation of single human hematopoietic stem cells capable of long-term multilineage engraftment. Science 333:218–221
5. Spangrude GJ, Heimfeld S, Weissman IL (1988) Purification and characterization of mouse hematopoietic stem cells. Science 241:58–62

6. Ogawa M, Matsuzaki Y, Nishikawa S, Hayashi S, Kunisada T et al (1991) Expression and function of c-kit in hemopoietic progenitor cells. J Exp Med 174:63–71
7. Okada S, Nakauchi H, Nagayoshi K, Nishikawa S, Miura Y et al (1992) In vivo and in vitro stem cell function of c-kit- and Sca-1-positive murine hematopoietic cells. Blood 80: 3044–3050
8. Boswell HS, Wade PM Jr, Quesenberry PJ (1984) Thy-1 antigen expression by murine high-proliferative capacity hematopoietic progenitor cells. I. Relation between sensitivity to depletion by Thy-1 antibody and stem cell generation potential. J Immunol 133: 2940–2949
9. Berman JW, Basch RS (1985) Thy-1 antigen expression by murine hematopoietic precursor cells. Exp Hematol 13:1152–1156
10. Williams DE, Boswell HS, Floyd AD, Broxmeyer HE (1985) Pluripotential hematopoietic stem cells in post-5-fluorouracil murine bone marrow express the Thy-1 antigen. J Immunol 135:1004–1011
11. Szilvassy SJ, Lansdorp PM, Humphries RK, Eaves AC, Eaves CJ (1989) Isolation in a single step of a highly enriched murine hematopoietic stem cell population with competitive long-term repopulating ability. Blood 74: 930–939
12. Ikuta K, Weissman IL (1992) Evidence that hematopoietic stem cells express mouse c-kit but do not depend on steel factor for their generation. Proc Natl Acad Sci USA 89: 1502–1506
13. Uchida N, Weissman IL (1992) Searching for hematopoietic stem cells: evidence that Thy-1.1lo Lin– Sca-1+ cells are the only stem cells in C57BL/Ka-Thy-1.1 bone marrow. J Exp Med 175:175–184
14. Spangrude GJ, Brooks DM (1992) Phenotypic analysis of mouse hematopoietic stem cells shows a Thy-1-negative subset. Blood 80: 1957–1964
15. Morrison SJ, Weissman IL (1994) The long-term repopulating subset of hematopoietic stem cells is deterministic and isolatable by phenotype. Immunity 1:661–673
16. Osawa M, Hanada K, Hamada H, Nakauchi H (1996) Long-term lymphohematopoietic reconstitution by a single CD34-low/negative hematopoietic stem cell. Science 273: 242–245
17. Adolfsson J, Borge OJ, Bryder D, Theilgaard-Monch K, Astrand-Grundstrom I et al (2001) Upregulation of Flt3 expression within the bone marrow Lin(–)Sca1(+)c-kit(+) stem cell compartment is accompanied by loss of self-renewal capacity. Immunity 15:659–669
18. Christensen JL, Weissman IL (2001) Flk-2 is a marker in hematopoietic stem cell differentiation: a simple method to isolate long-term stem cells. Proc Natl Acad Sci USA 98: 14541–14546
19. Yang L, Bryder D, Adolfsson J, Nygren J, Mansson R et al (2005) Identification of Lin(–)Sca1(+) kit(+)CD34(+)Flt3– short-term hematopoietic stem cells capable of rapidly reconstituting and rescuing myeloablated transplant recipients. Blood 105:2717–2723
20. Sitnicka E, Bryder D, Theilgaard-Monch K, Buza-Vidas N, Adolfsson J et al (2002) Key role of flt3 ligand in regulation of the common lymphoid progenitor but not in maintenance of the hematopoietic stem cell pool. Immunity 17:463–472
21. Adolfsson J, Mansson R, Buza-Vidas N, Hultquist A, Liuba K et al (2005) Identification of Flt3+ lympho-myeloid stem cells lacking erythro-megakaryocytic potential a revised road map for adult blood lineage commitment. Cell 121:295–306
22. Kiel MJ, Yilmaz OH, Iwashita T, Terhorst C, Morrison SJ (2005) SLAM family receptors distinguish hematopoietic stem and progenitor cells and reveal endothelial niches for stem cells. Cell 121:1109–1121
23. Balazs AB, Fabian AJ, Esmon CT, Mulligan RC (2006) Endothelial protein C receptor (CD201) explicitly identifies hematopoietic stem cells in murine bone marrow. Blood 107:2317–2321
24. Kent DG, Copley MR, Benz C, Wohrer S, Dykstra BJ et al (2009) Prospective isolation and molecular characterization of hematopoietic stem cells with durable self-renewal potential. Blood 113:6342–6350
25. Cheshier SH, Morrison SJ, Liao X, Weissman IL (1999) In vivo proliferation and cell cycle kinetics of long-term self-renewing hematopoietic stem cells. Proc Natl Acad Sci USA 96: 3120–3125
26. Foudi A, Hochedlinger K, Van Buren D, Schindler JW, Jaenisch R et al (2009) Analysis of histone 2B-GFP retention reveals slowly cycling hematopoietic stem cells. Nat Biotechnol 27:84–90
27. Ploemacher RE, Brons NH (1988) Cells with marrow and spleen repopulating ability and forming spleen colonies on day 16, 12, and 8 are sequentially ordered on the basis of increasing rhodamine 123 retention. J Cell Physiol 136:531–536
28. Ploemacher RE, Brons NH (1988) In vivo proliferative and differential properties of murine bone marrow cells separated on the basis of rhodamine-123 retention. Exp Hematol 16:903–907

29. Zijlmans JM, Visser JW, Kleiverda K, Kluin PM, Willemze R et al (1995) Modification of rhodamine staining allows identification of hematopoietic stem cells with preferential short-term or long-term bone marrow-repopulating ability. Proc Natl Acad Sci USA 92:8901–8905
30. Peng K, Visser AJ, van Hoek A, Wolfs CJ, Sanders JC et al (1990) Analysis of time-resolved fluorescence anisotropy in lipid-protein systems. I. Application to the lipid probe octadecyl rhodamine B in interaction with bacteriophage M13 coat protein incorporated in phospholipid bilayers. Eur Biophys J 18:277–283
31. Visser JW, de Vries P (1988) Isolation of spleen-colony forming cells (CFU-s) using wheat germ agglutinin and rhodamine 123 labeling. Blood Cells 14:369–384
32. Mulder AH, Visser JW (1987) Separation and functional analysis of bone marrow cells separated by rhodamine-123 fluorescence. Exp Hematol 15:99–104
33. Kim M, Cooper DD, Hayes SF, Spangrude GJ (1998) Rhodamine-123 staining in hematopoietic stem cells of young mice indicates mitochondrial activation rather than dye efflux. Blood 91:4106–4117
34. Goodell MA, Brose K, Paradis G, Conner AS, Mulligan RC (1996) Isolation and functional properties of murine hematopoietic stem cells that are replicating in vivo. J Exp Med 183:1797–1806
35. Goodell MA (2005) Stem cell identification and sorting using the Hoechst 33342 side population (SP). Curr Protoc Cytom, Chapter 9: Unit9 18
36. Morrison SJ, Hemmati HD, Wandycz AM, Weissman IL (1995) The purification and characterization of fetal liver hematopoietic stem cells. Proc Natl Acad Sci USA 92: 10302–10306
37. Harrison DE, Zhong RK, Jordan CT, Lemischka IR, Astle CM (1997) Relative to adult marrow, fetal liver repopulates nearly five times more effectively long-term than short-term. Exp Hematol 25:293–297
38. Jordan CT, McKearn JP, Lemischka IR (1990) Cellular and developmental properties of fetal hematopoietic stem cells. Cell 61:953–963
39. Jordan CT, Astle CM, Zawadzki J, Mackarehtschian K, Lemischka IR et al (1995) Long-term repopulating abilities of enriched fetal liver stem cells measured by competitive repopulation. Exp Hematol 23:1011–1015
40. Orlic D, Fischer R, Nishikawa S, Nienhuis AW, Bodine DM (1993) Purification and characterization of heterogeneous pluripotent hematopoietic stem cell populations expressing high levels of c-kit receptor. Blood 82: 762–770
41. Sanchez MJ, Holmes A, Miles C, Dzierzak E (1996) Characterization of the first definitive hematopoietic stem cells in the AGM and liver of the mouse embryo. Immunity 5:513–525
42. Yoder MC, Hiatt K, Dutt P, Mukherjee P, Bodine DM et al (1997) Characterization of definitive lymphohematopoietic stem cells in the day 9 murine yolk sac. Immunity 7: 335–344
43. Ito T, Tajima F, Ogawa M (2000) Developmental changes of CD34 expression by murine hematopoietic stem cells. Exp Hematol 28:1269–1273
44. Matsuoka S, Ebihara Y, Xu M, Ishii T, Sugiyama D et al (2001) CD34 expression on long-term repopulating hematopoietic stem cells changes during developmental stages. Blood 97: 419–425
45. Uchida N, Dykstra B, Lyons K, Leung F, Kristiansen M et al (2004) ABC transporter activities of murine hematopoietic stem cells vary according to their developmental and activation status. Blood 103:4487–4495
46. Kim I, He S, Yilmaz OH, Kiel MJ, Morrison SJ (2006) Enhanced purification of fetal liver hematopoietic stem cells using SLAM family receptors. Blood 108:737–744
47. Randall TD, Weissman IL (1997) Phenotypic and functional changes induced at the clonal level in hematopoietic stem cells after 5-fluorouracil treatment. Blood 89:3596–3606
48. Sato T, Laver JH, Ogawa M (1999) Reversible expression of CD34 by murine hematopoietic stem cells. Blood 94:2548–2554
49. Higuchi Y, Zeng H, Ogawa M (2003) CD38 expression by hematopoietic stem cells of newborn and juvenile mice. Leukemia 17:171–174
50. Spangrude GJ, Brooks DM, Tumas DB (1995) Long-term repopulation of irradiated mice with limiting numbers of purified hematopoietic stem cells: in vivo expansion of stem cell phenotype but not function. Blood 85: 1006–1016
51. Morrison SJ, Wandycz AM, Akashi K, Globerson A, Weissman IL (1996) The aging of hematopoietic stem cells. Nat Med 2: 1011–1016
52. Pearce DJ, Anjos-Afonso F, Ridler CM, Eddaoudi A, Bonnet D (2007) Age-dependent increase in side population distribution within hematopoiesis: implications for our understanding of the mechanism of aging. Stem Cells 25:828–835
53. Yilmaz OH, Kiel MJ, Morrison SJ (2006) SLAM family markers are conserved among hematopoietic stem cells from old and reconsti-

tuted mice and markedly increase their purity. Blood 107:924–930

54. Sudo K, Ema H, Morita Y, Nakauchi H (2000) Age-associated characteristics of murine hematopoietic stem cells. J Exp Med 192: 1273–1280
55. Kim M, Moon HB, Spangrude GJ (2003) Major age-related changes of mouse hematopoietic stem/progenitor cells. Ann N Y Acad Sci 996:195–208
56. Li CL, Johnson GR (1992) Rhodamine123 reveals heterogeneity within murine Lin−, Sca-1+ hemopoietic stem cells. J Exp Med 175:1443–1447
57. Sieburg HB, Cho RH, Dykstra B, Uchida N, Eaves CJ et al (2006) The hematopoietic stem compartment consists of a limited number of discrete stem cell subsets. Blood 107: 2311–2316
58. Muller-Sieburg CE, Cho RH, Karlsson L, Huang JF, Sieburg HB (2004) Myeloid-biased hematopoietic stem cells have extensive self-renewal capacity but generate diminished lymphoid progeny with impaired IL-7 responsiveness. Blood 103:4111–4118
59. Muller-Sieburg CE, Cho RH, Thoman M, Adkins B, Sieburg HB (2002) Deterministic regulation of hematopoietic stem cell self-renewal and differentiation. Blood 100: 1302–1309
60. Sieburg HB, Rezner BD, Muller-Sieburg CE (2011) Predicting clonal self-renewal and extinction of hematopoietic stem cells. Proc Natl Acad Sci USA 108:4370–4375
61. Beerman I, Bhattacharya D, Zandi S, Sigvardsson M, Weissman IL et al (2010) Functionally distinct hematopoietic stem cells modulate hematopoietic lineage potential during aging by a mechanism of clonal expansion. Proc Natl Acad Sci USA 107:5465–5470
62. Challen GA, Boles NC, Chambers SM, Goodell MA (2010) Distinct hematopoietic stem cell subtypes are differentially regulated by TGF-beta1. Cell Stem Cell 6:265–278
63. Morita Y, Ema H, Nakauchi H (2010) Heterogeneity and hierarchy within the most primitive hematopoietic stem cell compartment. J Exp Med 207:1173–1182
64. Matsuzaki Y, Kinjo K, Mulligan RC, Okano H (2004) Unexpectedly efficient homing capacity of purified murine hematopoietic stem cells. Immunity 20:87–93
65. Robinson SN, Seina SM, Gohr JC, Kuszynski CA, Sharp JG (2005) Evidence for a qualitative hierarchy within the Hoechst-33342 'side population' (SP) of murine bone marrow cells. Bone Marrow Transplant 35:807–818
66. Kiel MJ, Yilmaz OH, Morrison SJ (2008) CD150− cells are transiently reconstituting multipotent progenitors with little or no stem cell activity. Blood 111:4413–4414, author reply 4414–4415
67. Weksberg DC, Chambers SM, Boles NC, Goodell MA (2008) CD150− side population cells represent a functionally distinct population of long-term hematopoietic stem cells. Blood 111:2444–2451
68. Benveniste P, Frelin C, Janmohamed S, Barbara M, Herrington R et al (2010) Intermediate-term hematopoietic stem cells with extended but time-limited reconstitution potential. Cell Stem Cell 6:48–58

Chapter 2

Serial Transplantation of Bone Marrow to Test Self-renewal Capacity of Hematopoietic Stem Cells In Vivo

Charusheila Ramkumar, Rachel M. Gerstein, and Hong Zhang

Abstract

Hematopoietic stem cells (HSCs) have the ability to self-renew and replenish the blood and immune system for the life span of an individual. An age-associated decline in HSC function is responsible for the decreased immune function and increased incidence of myeloid diseases and anemia in the elderly. The changes in HSC function are thought to occur as the result of an intrinsic defect in the self-renewal potential of HSCs as they age. In this chapter, we describe a bone marrow serial transplantation protocol designed to test the self-renewal capacity of HSCs in vivo.

Keywords Hematopoietic stem cells, Serial transplantation, Stem cell self-renewal, Stem cell aging, Bone marrow

1 Introduction

Tissue-specific or adult stem cells are capable of self-renewal to preserve stem cell pools and differentiation into a variety of effector cells. With advancing age, the self-renewal capacity of stem cells invariably declines, eventually leading to the accumulation of unrepaired, damaged tissues in old organisms (1). Hematopoietic stem cells (HSCs), which give rise to various cellular components of blood, are known to exhibit a decline in self-renewal capacity with age. This is thought to contribute to decreased immune function in the elderly. The age-associated changes in HSC function have been extensively studied in mice, which include a strain-dependent increase in HSC number (2–5) and a decrease in their lymphoid differentiation potential (2, 4, 6). These changes are thought to occur due to an intrinsic defect in HSC self-renewal potential as they age (1).

Bone marrow transplantation is used to measure the "stemness" of HSCs, as a single HSC is sufficient to reconstitute the entire hematopoietic system of lethally irradiated recipients (7, 8). The fundamental principle of stem cell self-renewal is established by

Kursad Turksen (ed.), *Stem Cells and Aging: Methods and Protocols*, Methods in Molecular Biology, vol. 976, DOI 10.1007/978-1-62703-317-6_2, © Springer Science+Business Media, LLC 2013

the ability of bone marrow from such recipients to reconstitute secondary recipients with cells that are originated from the primary transplanted cells (7, 9). As a gold standard test for long-term self-renewal and multi-lineage potential of HSCs, serial transplantation of bone marrow cells is able to reconstitute lethally irradiated recipients in successive but limited transplants, reflecting the finite potential of HSC self-renewal (10–12). It has been shown that serial transplantation leads to a dose-dependent decrease in self-renewal capacity of HSCs (13–15), and this decline in stem cell function increases with the number of transplantations (10, 11, 16–19). Stem cell exhaustion has been reported in serial transfer experiments when unfractionated bone marrow (12, 18) or purified HSCs (20) are transplanted. The stem cell exhaustion in serial transplantation has been likened to an accelerated aging process, thus making it a powerful system to study HSC aging. Here we describe a protocol to test the long-term HSC (LT-HSC) self-renewal capacity using serial transplantation of unfractionated bone marrow.

2 Materials

2.1 Mouse Strains

1. Donor mice: Mice with the genotype of interest on a C57BL/6 background. Age-matched wild-type C57BL/6 littermates are used as controls.
2. Recipient mice: B6.SJL-*Ptprc*^{a} *Pepc*^{b}/BoyJ (CD45.1^{+}) aged 8–10 weeks (The Jackson Laboratory). This congenic strain with an allelic variant CD45.1 antigen can be distinguished from the C57BL/6 (CD45.2^{+}) donor by flow cytometric detection of the CD45 antigens. The use of CD45.1^{+} recipient mice allows the determination of the contribution of the donor cells (CD45.2^{+}) to reconstitution of bone marrow in lethally irradiated recipients, separate from residual recipient HSCs (CD45.1^{+}) that survive radiation and subsequently give rise to hematopoietic cells marked by CD45.1.

2.2 Antibiotics

1. Neomycin (200×): 5 g dissolved in 50 ml distilled water, and sterilized with a 0.45 μm filter.
2. Polymyxin-B (200×): 1 million units dissolved in 50 ml distilled water, and sterilized with a 0.45 μm filter.

2.3 For Irradiation

1. Cesium-137 radiation source.
2. Radiation chamber for mice.

2.4 For Harvesting Bone Marrow

1. Harvesting medium: Biotin, flavin and phenol red-deficient RPMI-1640 medium (Invitrogen) supplemented with 10 mM HEPES (pH 7.2), 1 mM EDTA, and 2% fetal bovine serum (FBS).

2. Viability staining solution (1,000× stock): 3 mg/ml of acridine orange and 5 mg/ml of ethidium bromide dissolved in distilled water. Prepare 100× solution with PBS.
3. 70% Ethanol.
4. Razor blades, dissecting scissors, and forceps.
5. 5 ml syringes with needles (25 G1/2 and 18 G).
6. 60-mm tissue culture plates and 15 ml Falcon tubes.
7. 70 μm nylon mesh (autoclaved).
8. Hemocytometer.
9. Temperature controlled centrifuge.
10. Fluorescence microscope.

2.5 For Injections

1. Isoflurane.
2. Sterile Dulbecco's phosphate buffered saline (PBS).
3. Insulin syringe with fitted needle (29 G1/2).
4. Mouse restrainer.
5. Heat lamp.

2.6 For Flow Cytometry Analysis

1. Antibodies: Lineage cocktail contains biotin-conjugated Ter119 (clone TER-119), CD11b (clone M1/70), Ly-6G (Gr1, clone RB6-8C5), CD45R (B220, clone RA3-6B2), CD19 (clone 1D3), and CD3e (clone 145-2C11). Additional antibodies for HSC analysis include Ly-6A/E (Sca1)-FITC (clone D7), CD117 (c-Kit)-PE-Cy7 (clone 2B8), CD135 (Flt3)-PE (clone A2F10), and CD150-APC (clone mShad150). Other materials include CD45.1-APC-eFluor 780 (clone A20), CD45.2-Alexa Fluor 700 (clone 104), streptavidin-eFluor 450, and Fc block CD16/CD32 (clone 2.4G2, from BioXCell). All except Fc block are purchased from eBioscience.
2. Staining medium: Biotin, flavin and phenol red-deficient RPMI-1640 medium (Invitrogen) supplemented with 10 mM HEPES (pH 7.2), 1 mM EDTA, 2% FBS, and 0.02% sodium azide.
3. 96-well flexible plates and 5 ml polystyrene tubes.
4. LSR II flow cytometry system with 5 lasers and 18 detectors (BD Biosciences).

3 Methods

3.1 Antibiotic Treatment

Recipient mice are treated with antibiotics in drinking water 24 h prior to exposure to radiation. Add 2 ml each of 200× antibiotic stock solutions to 396 ml of autoclaved acidified water. Drinking water with antibiotics must be changed twice weekly until 1 month after transplantation (see Note 1).

3.2 Irradiation

Recipient mice are exposed to a lethal dose of 10 Gy (1,000 Rads) whole body radiation using a Cesium-137 source (see Note 2). At least five recipient mice are needed for each donor. For later (>3) cycles of transplantation, at least ten recipient mice are used per donor.

3.3 Harvesting Donor Bone Marrow

Harvest bone marrow on ice in a laminar flow hood. Using sterile techniques is essential while flushing and preparing bone marrow for injection.

1. Euthanize the donor mouse with isoflurane and cervical dislocation, and immerse the mouse in 70% Ethanol completely.
2. Cut the hind limbs away from the hip joint. Be careful not to break the femur while dissecting the hip joint. Similarly, cut the forelimbs away from the shoulder joint. Place dissected limbs in a 60-mm plate with cold bone marrow harvesting medium on ice.
3. Hold one dissected limb with a pair of forceps, and scrape away skin and muscle with a razor blade until only bone remains. Try to get rid of as much tissue that is attached to the bone as possible. Repeat this procedure with all limbs.
4. In a separate 60-mm plate with cold bone marrow harvesting medium on ice, disarticulate the knee joint by cutting through it with a razor blade. Cut tibia, femur, and humerus bones at both ends so that marrow cavities are open.
5. Fill a 5 ml syringe with bone marrow harvesting medium and fit a 25 G1/2 needle on the syringe. Hold the bone with a pair of forceps. Fit the needle into one of the cut ends of the bone and flush the bone marrow out. Repeat flushing until the color of the bone changes from a pinkish tinge to almost completely white. Repeat this procedure with all bones.
6. When all marrow has been flushed out, change the 25 G1/2 needle to an 18 G needle. The marrow is in large chunks and can be broken up into a single cell suspension by passing it through an 18 G needle several times.
7. Once a single cell suspension has been achieved, filter these cells through a 70 μm nylon mesh into a 15 ml Falcon tube. Bring the final volume of cells to 10 ml with staining medium. Place cells on ice.

3.4 Cell Counting

1. Mix 10 μl of cell suspension, 89 μl of staining medium, and 1 μl of 100× viability stain solution by pipetting. Add 10 μl of this mixture in the hemocytometer.
2. Count green (viable) cells in a 5 × 5 grid under fluorescence microscope. Red/orange cells are dead (ethidium bromide-positive) and can be counted in order to determine the ratio of live to dead cells in a sample.

3. Calculate concentration and total number of live cells:

$$\text{Cells / ml in hemocytometer} \\ = \#\ \text{green cells in the } 5 \times 5 \text{ square} \times 10^4$$

$$\text{Cells / ml in tube} = \text{cells / ml in hemocytometer} \times 10$$

$$\text{Total live cells in tube} = \text{cells / ml in tube} \times 10$$

3.5 Preparation of Bone Marrow Cells for Injection

1. 2×10^6 bone marrow cells (~200 HSCs) per recipient mouse are usually used. Calculate the volume of cells sufficient for required injections plus two extra injections (see Note 3).
2. Spin down cells at 1,500 rpm ($388 \times g$) for 5 min in a centrifuge prechilled to 4°C. Resuspend cell pellet in 10 ml sterile Dulbecco's PBS.
3. Spin down cells again and resuspend cells in Dulbecco's PBS at a concentration of 2×10^6 cells per 200 μl. Place resuspended cells on ice and bring them to mouse facility for injection.
4. The remaining cells are centrifuged and resuspended in staining medium at 6×10^7 cells/ml. They are used to stain for the LT-HSC population in flow cytometry analysis (see Subheading 3.7).

3.6 Bone Marrow Cells Injection

Tail vein or retro-orbital injection can be used to inject donor bone marrow cells into recipient mice. Tail vein injection is commonly used, but it can be difficult to visualize tail veins in C57BL/6 mice. Alternatively, cells are injected into the retro-orbital sinus of the mouse in retro-orbital injection. While this method is technically less challenging, it has the limitation of being able to inject a maximum volume of 200 μl into the sinus. In addition, if the needle scrapes the cornea while injecting, the chances of developing corneal ulcers are high. Both methods of injection require practice, and we recommend practicing injection with PBS a few days before the actual experiment.

3.6.1 Tail Vein Injection

1. Prepare the sample for injection by filling 200 μl of cell suspension into an insulin syringe and keep aside (see Note 4).
2. Place an irradiated recipient mouse in a mouse restrainer. Warm the tail by shining a heat lamp on the tail briefly. This causes vasodilation and enables easy visualization of the tail veins.
3. Hold the tail in one hand, and select the vein you want to inject. There are two veins, lateral and medial, in each tail. Insert the tip of the needle into the vein, and withdraw slightly. If blood is drawn into the syringe, the needle is in the vein. Inject the cells quickly (see Note 5).

3.6.2 Retro-orbital Injection

1. Anesthetize the recipient mouse with isoflurane as follows. Soak a nestlet with isoflurane and place it in an empty cage. Place the mouse in this cage and wait until its respiration slows. This usually takes about 30 s.
2. Once respiration has slowed, take the mouse out and open an eye wide by spreading the lids with one hand. The retro-orbital sinus will be visible as a small opening at the medial corner of the eye.
3. Place the tip of the needle in the sinus, hold the syringe at an angle of 45° to the eye and inject cells into the sinus (see Note 6).
4. Place the animal back in its cage and observe until it awakens from the anesthesia completely.

3.7 Fluorescence-Activated Cell Sorting Analysis of LT-HSCs in Bone Marrow

1. Cells resuspended at 6×10^7/ml in staining medium are incubated with anti-CD16/CD32 antibody at 1 μg/10^6 cells for 10 min on ice to block the Fc receptors.
2. 25 μl of these cells are then incubated with 25 μl of each primary antibody in staining medium for 20 min in a 96-well flexible plate on ice.
3. Spin down cells at 1,500 rpm (388 × *g*) for 5 min in a centrifuge prechilled to 4°C, and resuspend cell pellets in 100 μl of staining medium. Repeat the washing step two more times.
4. Cells stained with biotin-labeled antibodies (lineage cocktail) are incubated with streptavidin-eFluor 450 for 15 min on ice and washed three times with staining medium.
5. After the final wash, cells are resuspended in 1 μg/ml propidium iodide (PI) in staining medium for the exclusion of dead cells.
6. Flow cytometry analysis is performed on a 5-laser, 18-detector LSR II fluorescence-activated cell sorting (FACS) machine using 405, 488, 561, and 633-nm lasers. Data are analyzed using FlowJo software (Treestar). A representative experiment in Fig. 1 shows the progressive gating strategy used to analyze LT-HSCs from live bone marrow cells (PI-negative). Lineage negative (Lin^-) cells are those lacking significant expression of Gr-1, CD11b, Ter119, CD3, B220, and CD19. Long-term HSCs (LT-HSCs), short-term HSCs (ST-HSCs), and multipotent progenitors (MPP) are characterized by $Lin^-Sca1^+c\text{-}kit^{++}CD150^+Flt3^-$, $Lin^-Sca1^+c\text{-}kit^{++}CD150^-Flt3^-$, and $Lin^-Sca1^+c\text{-}kit^{++}CD150^-Flt3^+$, respectively (21, 22).

3.8 Monitoring Mice and Subsequent Transplantations

1. After transplantation, recipient mice must be monitored daily for signs of ill-health including pallor, ruffled fur and lethargy. Typically, recipient mice receiving inadequate injections will become increasingly pale and sick, and die between 10 days and 2 weeks after irradiation due to bone marrow failure.

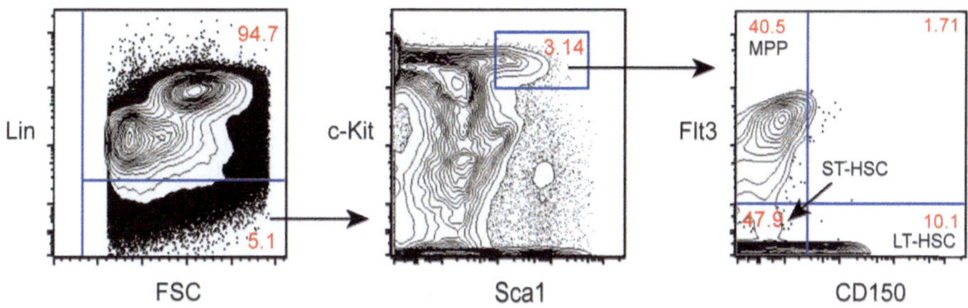

Fig. 1 A representative FACS analysis of LT-HSCs in bone marrow of a wild-type C57BL/6 mouse. Gates used for FACS analysis are displayed as *blue boxes*. The frequency of each gated population as a percent of the displayed cells is shown in *red*. Only live cells (propidium iodide excluding) are displayed. Lineage negative (Lin^-) cells (*left panel*) are gated and then displayed in the *middle panel*. LSK ($Lin^-Sca1^+c\text{-}Kit^{++}$, *middle panel*) cells are gated and displayed in the *right panel*

2. When 2 months after the injection have elapsed, the stem cell numbers reach homeostasis and secondary transplants can be performed. These recipients can be used as donors for the next transplantation. With every ensuing transplant, a sequential decrease in the frequency of the LT-HSC population is expected. HSCs from wild-type mice normally can reconstitute recipient bone marrow for 4–5 cycles of transplantation before stem cell exhaustion occurs.
3. Relative contributions of the donor and residual recipient bone marrow to the reconstitution in recipient mice can be determined by staining for CD45.1 and CD45.2 of bone marrow from the first transplantation onward. More than 90% of the cells are usually derived from the donor (CD45.2) bone marrow.

4 Notes

1. Leave a note on cages of recipient mice that they are on antibiotic treatment, so animal care technicians do not change the water.
2. Set up irradiation late in the afternoon, and irradiated mice can be transplanted with fresh bone marrow the next morning. Make sure to check the Cesium-137 source to calculate the radiation dosage every time you irradiate.
3. If injecting five mice, calculate the volume of cells required for 14×10^6 cells (5 plus 2 extra injections, 2×10^6 cells/injection).
4. Before injection, make sure that the cells are not chilled by warming them between your hands briefly.
5. Watch for clearing of the vein lumen; this indicates a successful injection. If you feel any resistance while injecting, it means the

needle has slipped out of the vein. If that happens, withdraw the syringe and reinsert it into the vein proximal to the original injection site.

6. If cells regurgitate back, that means they are not in the sinus. Take out the needle and reinsert. If cells come out through the nose, the procedure is unsuccessful and must be repeated. This process has to be performed quickly, as the effect of isoflurane wears off within 1–2 min.

References

1. Rossi DJ, Jamieson CH, Weissman IL (2008) Stems cells and the pathways to aging and cancer. Cell 132(4):681–696
2. Rossi DJ et al (2005) Cell intrinsic alterations underlie hematopoietic stem cell aging. Proc Natl Acad Sci USA 102(26):9194–9199
3. Morrison SJ, Wandycz AM, Akashi K, Globerson A, Weissman IL (1996) The aging of hematopoietic stem cells. Nat Med 2(9):1011–1016
4. Sudo K, Ema H, Morita Y, Nakauchi H (2000) Age-associated characteristics of murine hematopoietic stem cells. J Exp Med 192(9): 1273–1280
5. Chambers SM, Goodell MA (2007) Hematopoietic stem cell aging: wrinkles in stem cell potential. Stem Cell Rev 3(3):201–211
6. Liang Y, Van Zant G, Szilvassy SJ (2005) Effects of aging on the homing and engraftment of murine hematopoietic stem and progenitor cells. Blood 106(4):1479–1487
7. Wu AM, Till JE, Siminovitch L, McCulloch EA (1968) Cytological evidence for a relationship between normal hemotopoietic colony-forming cells and cells of the lymphoid system. J Exp Med 127(3):455–464
8. Abramson S, Miller RG, Phillips RA (1977) The identification in adult bone marrow of pluripotent and restricted stem cells of the myeloid and lymphoid systems. J Exp Med 145(6):1567–1579
9. Keller G, Snodgrass R (1990) Life span of multipotential hematopoietic stem cells in vivo. J Exp Med 171(5):1407–1418
10. Harrison DE, Astle CM, Delaittre JA (1978) Loss of proliferative capacity in immunohemopoietic stem cells caused by serial transplantation rather than aging. J Exp Med 147(5): 1526–1531
11. Harrison DE, Stone M, Astle CM (1990) Effects of transplantation on the primitive immunohematopoietic stem cell. J Exp Med 172(2):431–437
12. Harrison DE, Astle CM (1982) Loss of stem cell repopulating ability upon transplantation. Effects of donor age, cell number, and transplantation procedure. J Exp Med 156(6): 1767–1779
13. Iscove NN, Nawa K (1997) Hematopoietic stem cells expand during serial transplantation in vivo without apparent exhaustion. Curr Biol 7(10):805–808
14. Mauch P, Hellman S (1989) Loss of hematopoietic stem cell self-renewal after bone marrow transplantation. Blood 74(2):872–875
15. Hellman S, Mauch P (1984) Implications of a proliferative limitation on hematopoietic stem cells. Prog Clin Biol Res 148:51–58
16. Siminovitch L, Till JE, McCulloch EA (1964) Decline in colony-forming ability of marrow cells subjected to serial transplantation into irradiated mice. J Cell Physiol 64:23–31
17. Cudkowicz G, Upton AC, Shearer GM, Hughes WL (1964) Lymphocyte content and proliferative capacity of serially transplanted mouse bone marrow. Nature 201:165–167
18. Ross EA, Anderson N, Micklem HS (1982) Serial depletion and regeneration of the murine hematopoietic system. Implications for hematopoietic organization and the study of cellular aging. J Exp Med 155(2):432–444
19. Micklem HS, Ross E (1978) Heterogeneity and ageing of haematopoietic stem cells. Ann Immunol 129(2–3):367–376
20. Spangrude GJ, Brooks DM, Tumas DB (1995) Long-term repopulation of irradiated mice with limiting numbers of purified hematopoietic stem cells: in vivo expansion of stem cell phenotype but not function. Blood 85(4):1006–1016
21. Yilmaz OH, Kiel MJ, Morrison SJ (2006) SLAM family markers are conserved among hematopoietic stem cells from old and reconstituted mice and markedly increase their purity. Blood 107(3):924–930
22. Kiel MJ et al (2005) SLAM family receptors distinguish hematopoietic stem and progenitor cells and reveal endothelial niches for stem cells. Cell 121(7):1109–1121

Chapter 3

Mouse Hematopoietic Stem Cell Transplantation

Hui Cheng, Paulina H. Liang, and Tao Cheng

Abstract

Hematopoietic stem cells (HSCs) are capable of self-renewal and multi-lineage reconstitution of hematopoiesis in irradiated transplant recipient mice. As such, bone marrow transplantation (BMT) is a major assay commonly used to examine murine HSC activity. BMT traditionally involves injection of HSCs into lethally irradiated recipients via the tail vein, then subsequently analyzing donor engraftment. Here, we describe the methods for assaying HSC reconstitution in direct, competitive, and serial BMT.

Keywords Hematopoietic stem cell (HSC), Bone marrow transplantation (BMT), Competitive BMT (cBMT), Serial BMT (sBMT), Single cell transplantation

1 Introduction

As one of the best studied tissue stem cell types, the hematopoietic stem cell (HSC) is capable of self-renewal and gives rise to all blood and immune cells in the body (1, 2). In mice of different genetic backgrounds, the activity of HSCs appears to be correlated with lifespan (3, 4), although studies have shown that HSC functionality could be sustained even longer (5). Whether HSCs become senescent *in vivo* under homeostatic conditions is still a subject of debate; however, it is clear that the functions of HSCs do change during the lifetime of an organism. One of the main assays used to assess the potential activity of hematopoietic cells is the *in vivo* transplantation of HSCs into lethally irradiated recipients (6, 7). This *in vivo* assay remains a gold-standard in defining HSCs and their functional potential.

Antibody-based sub-selection of HSCs combined with long-term transplantation has enabled a detailed understanding of the commitment and differentiation of HSCs into progenitor cells and ultimately mature blood cells (8). There are many variations of long-term repopulating assays, the most common of which is the competitive repopulation assay (9). This assay measures the functional potential of HSCs from an experimental donor group against

Kursad Turksen (ed.), *Stem Cells and Aging: Methods and Protocols*, Methods in Molecular Biology, vol. 976,
DOI 10.1007/978-1-62703-317-6_3,

a set number of control HSCs (usually whole bone marrow cells from congenic wild-type mice). The most immature HSC is capable of sustaining hematopoiesis throughout serial transplantation (10, 11) in which HSCs are transplanted into sequential transplant recipients, and the ability of the cells to reconstitute hematopoiesis is determined. More recently, single cell transplantation has been used to assess multi-lineage reconstituting ability of a single HSC (12). Here, we provide a brief overview of the methods commonly used for murine HSC transplantation.

2 Materials

2.1 Mice

In general, female C57BL/6J (B6) congenic mice are used for transplantation assays. For donor versus host choice, we routinely use B6-Ly5.1 mice as donors and B6-Ly5.2 mice as hosts. Ly5.1 and Ly5.2 epitopes can be recognized by anti-CD45.1 and anti-CD45.2 antibodies. Mice aged 8–10 weeks are usually used as recipients.

2.2 Buffers and Reagents

1. Phosphate-buffered saline (PBS): Dissolve the following in 800 ml of H_2O: 8 g NaCl, 0.2 g KCl, 1.44 g Na_2HPO_4, and 0.24 g KH_2PO_4. Adjust pH to 7.2–7.4 with HCl. Add additional water to make a final volume of 1 l, then autoclave to sterilize.
2. BSA (Sigma-Aldrich) or FBS (Gibco).
3. Staining buffer (*PBE*): PBS, 0.5% BSA or 2% heat-inactivated FBS and 2 mM EDTA. Store at 4°C.
4. DAPI (Sigma-Aldrich): 1 mg/ml in distilled H_2O stock solution. Store at −20°C.
5. Erythrocyte lysis buffer: 10 mM $KHCO_3$, 150 mM NH_4Cl, and 0.1 mM EDTA (pH 8.0).
6. 0.4% Trypan blue solution (Sigma-Aldrich).
7. Medium: IMDM (Gibco).

2.3 Antibodies and Conjugation (see Note 1)

All antibodies are from BD Biosciences or e-Bioscience unless specified otherwise.

1. Linage markers: Biotin or PE-Cy7 conjugated anti-CD3 (145-2C11), anti-CD4 (GK1.5 or RM4-5), anti-CD8 (53-6.7), anti-CD45R (RA3-6B2), anti-CD11b (M1/70), anti-Gr-1(RB6-8C5), and anti-TER-119 (TER-119).
2. PE or PE-Cy7 anti-Sca-1 (D7).
3. APC anti-c-Kit (2B8).
4. FITC anti-CD34 (RAM34).
5. PE anti-CD150 (TC15-12F12.2, Biolegend).

6. Biotin conjugated CD48 (HM48-1).
7. Streptavidin APC-Cy7.
8. c-Kit (CD117) magnetic beads (Miltenyi Biotec).

2.4 Equipment

1. MACS MultiStand (Miltenyi Biotec).
2. MACS separator for MS or LS column (Miltenyi Biotec).
3. MS or LS column (Miltenyi Biotec).
4. 5 ml polystyrene round-bottom tube (BD Falcon).
5. Round-bottomed 96-well micro-titer plate (Corning) or equivalent.
6. 15 ml Centrifuge tube (Corning) or equivalent.
7. 1 ml or 5 ml syringe.
8. 30–70 μm nylon mesh.

2.5 Flow Cytometer

An analytical cytometer and cell sorter with multiple-laser excitation is required for this protocol.

1. For analysis, a BD LSRII flow cytometer equipped with a 488 nm laser (6 color octagon) and a 633 nm laser (3 color trigon) is used.
2. For sorting, a BD FACSAriaII sorter equipped with a 355 nm UV laser (2 color trigon), a 488 nm laser (5 color octagon), and a 633 nm laser (2 color trigon) is used.

2.6 Irradiator

For irradiation of recipient mice, a ^{137}Csγ-irradiator or X-ray irradiator is required.

3 Methods

3.1 Irradiate Recipient Mice

Irradiate Ly5.2 recipient mice at a lethal dose (9.5–11 Gy is commonly used by different labs) 5–12 h prior to BM transplantation.

3.2 Isolation of Bone Marrow Cells from Donor Mice

1. Sacrifice the Ly5.1 donor mice according to procedures commonly used and approved by the local institution (e.g., CO_2 inhalation).
2. Dissect out the tibia and femur from both legs and put them in a 6 mm or 10 mm culture dish containing ice-cold PBS. Using sharp surgical scissors, remove the muscle from the bones.
3. For each bone, take up 3 ml ice-cold PBE using a 5 ml syringe and a 25-gauge needle. Insert needle into one end of the bone and flush out the BM into a 5 ml tube.
4. Thoroughly mix the cell suspension and pass the cells through a 30–70 μm nylon mesh filter into a new 5 ml tube to remove cell clumps.

5. Count the number of nucleated cells with a hemocytometer or an auto-counter.
6. Keep cell suspension on ice until use.

3.3 c-Kit (CD117) Positive-Selection of BM Cells Using MicroBeads (see Note 2)

1. Centrifuge the cell suspension at 300 × *g* for 10 min at 4°C and discard supernatant.
2. Resuspend cell pellet in 80 μl of PBE per 10^8 total cells.
3. Add 20 μl of CD117 MicroBeads per 10^8 total cells.
4. Mix well and incubate for 15 min on ice.
5. Wash cells by adding 2 ml of PBE per 10^8 cells and centrifuge at 300 × *g* for 10 min. Discard supernatant.
6. Resuspend up to 10^8 cells in 500 μl of PBE.
7. Place MS or LS column in the magnetic field of a suitable MACS Separator.
8. Prepare column by rinsing with appropriate volume of PBE. (MS: 500 μl, LS: 3 ml)
9. Transfer cell suspension onto the column.
10. Collect unlabeled cells that pass through and wash column with PBE. Perform washing steps by adding PBE three times. (MS: 3 × 500 μl, LS: 3 × 3 ml)
11. Remove column from the separator and place it into a 15 ml tube.
12. Pipette an appropriate amount of PBE onto column. Immediately flush out the magnetically labeled cells by firmly pushing the plunger into the column. (MS: 1 ml, LS: 5 ml)
13. Count the number of nucleated cells and spin down the cell suspension at 500 × *g* for 10 min at 4°C.

3.4 Antibody Staining and Flow Cytometry

1. Adjust the cell concentration to about 1×10^7 cells per 50 μl with PBE (see Note 3).
2. For LSK or $CD34^-$LSK cell sorting, add the following antibodies to the cells per 50 μl staining system, and incubate cells on ice for 60–90 min (see Note 4):

   FITC CD34: 4 μl

   PE Sca-1: 1 μl

   APC c-Kit: 2 μl

   PE-Cy7 Lineage cocktail: 4.75 μl (Table 1)
3. For $CD34^-$LSK $CD150^+CD48^-$ cell sorting, add 2 μl Biotin lineage cocktail (Table 1) with 0.5 μl Biotin CD48 antibody. Mix well and incubate cells on ice for 30 min.
4. Wash cells once with 2 ml PBE and resuspend cell pellet in 50 μl PBE.

Table 1
Preparation of a cocktail of Lineage antibodies

	Biotin	PE-Cy7
Ter119	10 μl	10 μl
Gr-1	10 μl	2.5 μl
Mac-1	5 μl	10 μl
B220	5 μl	10 μl
CD3	5 μl	10 μl
CD4	2.5 μl	2.5 μl
CD8	2.5 μl	2.5 μl
Total	40 μl	47.5 μl
Cocktail[a]	2 μl/10^7 cells	4.75 μl/10^7 cells

[a]10^7 cells in 50 μl staining buffer
The volume of each antibody to make up the cocktail may differ from individual labs

5. Add the following antibodies to the cells per 50 μl staining system and incubate cells on ice for 60–90 min:

   FITC CD34: 4 μl

   PE-Cy7 Sca-1: 1 μl

   APC c-Kit: 2 μl

   PE CD150: 1 μl

   Streptavidin APC-Cy7: 1.5 μl
6. After incubation, wash cells of step 2 or 5 twice with 2 ml PBE.
7. Filter cells through a 30–70 μm nylon mesh.
8. Suspend the cells in the appropriate volume of ice-cold PBE for flow analysis.
9. For discrimination of dead cells, add DAPI at a final concentration of 1 μg/ml immediately before running samples for flow cytometry (see Note 5) (Fig. 1).

3.5 Reconstitution: BM Transplantation and Serial BMT

1. Adjust the Ly5.1 BM cell concentration to 2×10^6 cells per ml with PBS.
2. Inject 1×10^6 cells into each recipient via the tail vein.
3. Collect the peripheral blood of recipient mice every 4 weeks to monitor the reconstitution.
4. After long-term engraftment (>16 weeks) is established in the primary recipients, sacrifice the mice and harvest the BM.

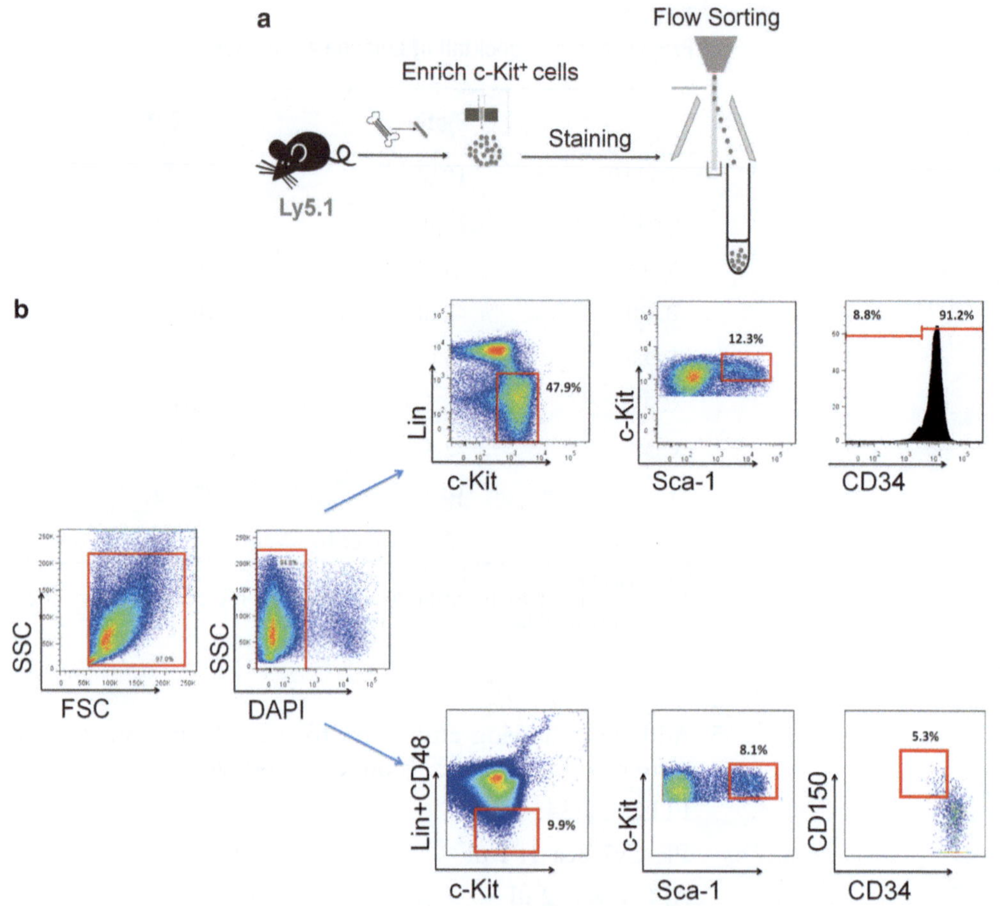

Fig. 1 Scheme of HSC sorting. (**a**) BM cells from Ly5.1 mice are enriched using MACS. After staining with the antibodies, target cells are sorted by FACS. (**b**) FACS plots show the gating strategy for HSC sorting. HSCs are highly enriched in the Lin$^-$Sca1$^+$c-Kit$^+$CD34$^-$ or Lin$^-$Sca1$^+$c-Kit$^+$CD34$^-$CD150$^+$CD48$^-$ fraction of the BM

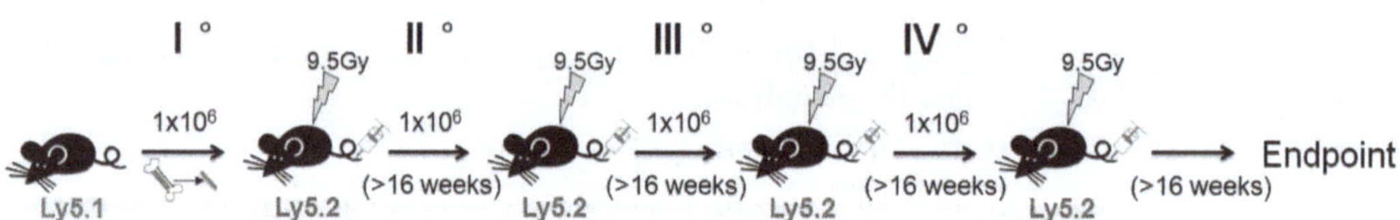

Fig. 2 Scheme of primary and serial bone marrow transplantation. BM cells are collected from Ly5.1 mice and transplanted into lethally irradiated (9.5 Gy) Ly5.2 recipients (1×10^6 cells/recipient). After long-term engraftment is established in the primary recipients, the same dose of BM cells are pooled from the primary recipients and re-transplanted into lethally irradiated Ly5.2 secondary recipients. The same procedure is repeated an additional three times. I°, II°, III°, and IV° indicate first, second, third, and fourth sBMTs, respectively

The same dose of BM cells are pooled from the primary recipients and re-transplanted into lethally irradiated secondary recipients. The same procedure is commonly repeated three times, totaling four transplants (see Note 6) (Fig. 2).

3.6 Competitive Bone Marrow Transplantation Coupled with Serial Transfer

Donor cells are usually different according to the experimental design. Briefly, whole BM cells and purified HSCs are commonly used.

1. For the primary competitive bone marrow transplantation (cBMT), 5×10^5 Ly5.1 BM cells or 500 Ly5.1 LSKs (50 $CD34^-$LSKs) are transplanted into lethally irradiated Ly5.2 recipient mice in competition with 5×10^5 Ly5.2 BM cells.
2. Peripheral blood of recipient mice is collected every 4 weeks after transplantation. The relative contribution of donor cells (Ly5.1) and competitive cells (Ly5.2) in reconstituted recipients is measured by flow cytometry.
3. For serial transplants, recipient mice are sacrificed 16–24 weeks after primary transplantation. Two different methods are available.

   Method 1: BM cells are pooled together (without sorting) and 1×10^6 cells are injected into lethally irradiated Ly5.2 recipients. Subsequent transplantations are performed in the same manner.

   Method 2: BM cells collected from primary recipients are pooled together, and then Ly5.1 BM cells or Ly5.1 HSCs are flow purified. 5×10^5 Ly5.1 BM cells or 500 Ly5.1 LSK (50 $CD34^-$LSK) are transplanted into lethally irradiated Ly5.2 recipients together with freshly isolated 5×10^5 Ly5.2 BM cells. Subsequent transplantations are performed in the same manner (see Note 7) (Fig. 3).

3.7 Single Cell Transplantation

1. Directly sort $CD34^-$LSK $CD150^+CD48^-$ (see Note 8) cells from Ly5.1 BM at one cell per well into a round-bottom 96-well micro-titer plate. Each well contains 150 μl IMDM plus 10% FBS.
2. Place the plate in a 37°C incubator for 1–2 h so that the cells sink down to the well bottoms, permitting easier identification under an inverted microscope.
3. Visually verify under an inverted microscope that one cell is present per well.
4. To each well where one cell is found, add 100 μl PBS containing 2×10^5 Ly5.2 BM cells (see Note 9).
5. Take up all medium from one well with a 1 ml syringe and inject into a lethally irradiated recipient.
6. Monitor the engraftment at 4 and 16 weeks after transplantation (Fig. 4).

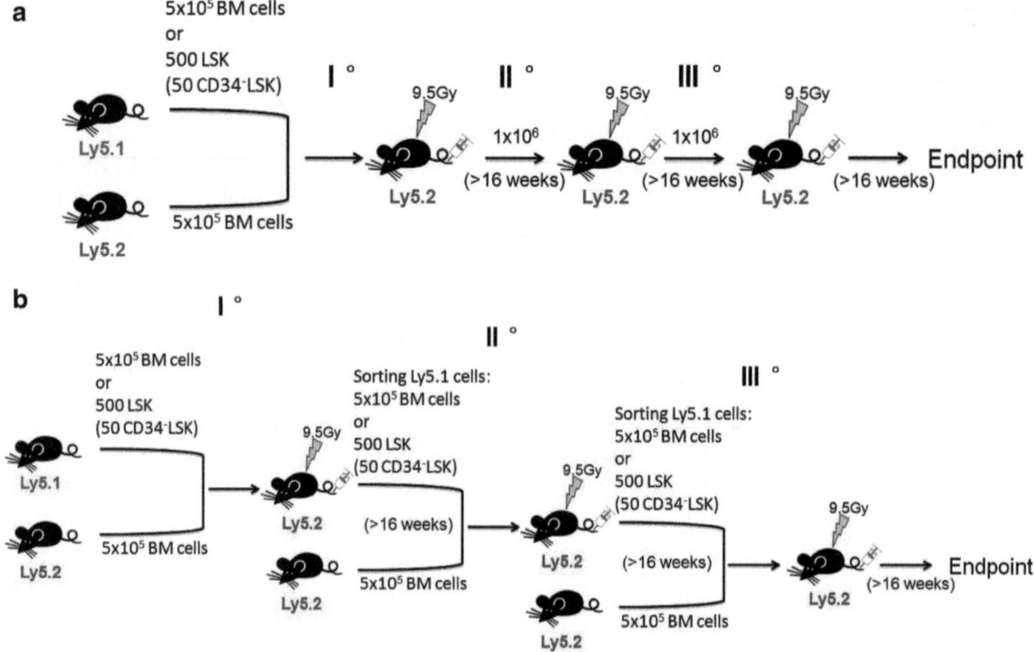

Fig. 3 Scheme of cBMT coupled with serial transfer. For primary cBMT, lethally irradiated Ly5.2 mice were reconstituted with 5×10^5 BM cells or purified HSCs (500 LSK or 50 CD34⁻LSK) from Ly5.1 mice, in competition with 5×10^5 BM cells from Ly5.2 mice. For the serial transplantation analysis, two different methods are available as described in the context

4 Notes

1. Antibodies for flow cytometry are available from numerous suppliers. Most of the antibodies described in this protocol are from BD bioscience, Ebiosciences, and Biolegend. The antibody clones used in this protocol are selected based on the publications describing their use in identifying the bone marrow HSC. The panel of antibody–fluorochrome combinations is optimized according to our experience, and other combinations are readily available from different laboratories.
2. The procedures described here are almost according to the manufacturer's protocol. However, we reduced the amount of Microbeads for staining as we have found that 20 μl of Microbeads is sufficient for one adult B6 mouse. This step reduces the time required for cell sorting as the primitive cells are greatly enriched by removing the majority of the c-Kit⁻ cells and it also helps to increase purity of the sorted target populations.
3. Sometimes this step may be not necessary. To stain cells with an antibody, the concentration of the antibody is the key,

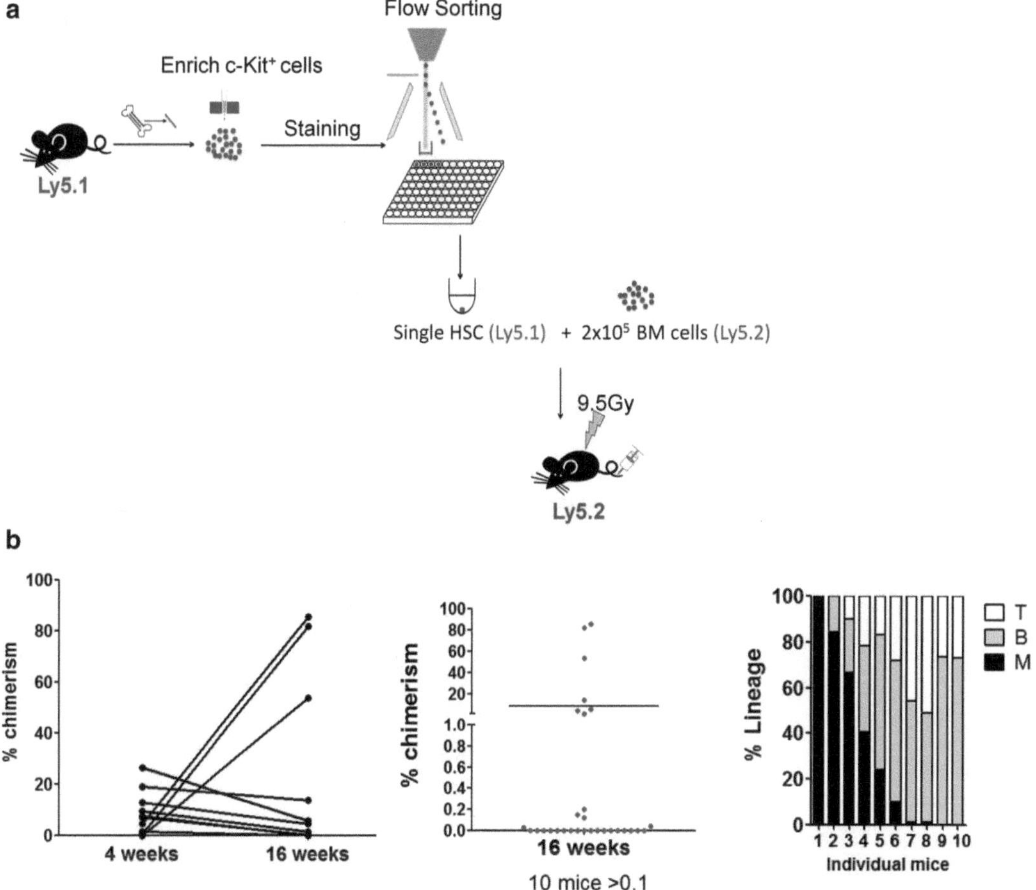

Fig. 4 Scheme of single cell transplantation and representative results. (**a**) Single HSCs (Lin$^-$Sca1$^+$c-Kit$^+$CD34$^-$CD150$^+$CD48$^-$) are directly sorted into a 96-well plate, at one cell per well, then mixed with 2×10^5 Ly5.2 BM cells. Cells are injected into lethally irradiated (9.5 Gy) Ly5.2 recipients. (**b**) Long-term reconstitution with single HSCs. Recipient mice were analyzed 16 weeks after transplantation. In this experiment, we considered reconstitution to be achieved when the percentage of donor (Ly5.1) chimerism was 0.1% or more at 16 weeks after transplantation, regardless of which lineage was reconstituted. A total of 30 recipient mice were analyzed. Among them, 10 mice showed long-term reconstitution and 8 mice showed long-term multi-lineage reconstitution

not the cell concentration. However, we usually adjust the cell concentration to 0.5×10^7–1.5×10^7 per 50 μl staining buffer to get comparable results.

4. Antibody staining on ice or at 4°C is recommended, as higher temperatures may lead to nonspecific cell labeling. A minimum of 60 min is recommended for good staining of cells with FITC anti-CD34 antibody, although 30 min is enough for the other antibodies (12). The final concentration for each individual antibody during the staining process could be different because of the source of the antibodies. We always get good results using the volume of antibodies described in this protocol.

5. Because of manipulations before cell sorting, there are always a certain percentage of dead cells after staining. To obtain the most accurate results, dead cells should be discriminated by DAPI or other DNA dyes such as PI and 7-AAD.
6. In general, 12-week reconstitution is considered to be from short-term HSC contribution. Therefore, to monitor the long-term reconstitution post-transplant, a minimum of 16 weeks is required and an optimal 6 months is suggested (8). To assess the self-renewal capacity of HSCs, serial transplantation is necessary. Usually, HSCs can be transplanted four to five times before exhaustion (13).
7. The competitor cells from unmanipulated fresh BM are repetitively used in each sequential cBMT as a standard measure for determining the repopulation ability of the original input cells over the course of the serial transfer (14).
8. The most purified HSC is required for single cell assays. HSC purity from young adult mice can be enhanced using the SLAM family markers, specifically CD150 and CD48. 47% of single $CD150^{+}CD48^{-}Sca\text{-}1^{+}Lineage^{-}c\text{-}Kit^{+}$ bone marrow cells (1 in 2.1) were capable of long-term multi-lineage reconstitution in irradiated mice (15).
9. Usually, 2×10^5 competitor cells (Whole BM cells) are used to support the short-term survival of lethally irradiated recipients. Sometimes, to increase the engraftment of a single HSC, competitor cells are depleted of Sca-1 positive BM cells using MACS.

Acknowledgments

We thank Dr. Hongmei Shen for her critical review of this manuscript. This work was supported by the grants from the Ministry of Science and Technology of China (2011CB964800) and the National Science Foundation of China (81090410 and 30825017) to TC. TC was a recipient of the Scholar Award from the Leukemia & Lymphoma Society (1027–08).

References

1. Abramson S, Miller RG, Phillips RA (1977) The identification in adult bone marrow of pluripotent and restricted stem cells of the myeloid and lymphoid systems. J Exp Med 145:1567–1579
2. Lemischka IR, Raulet DH, Mulligan RC (1986) Developmental potential and dynamic behavior of hematopoietic stem cells. Cell 45:917–927
3. Geiger H, Van Zant G (2002) The aging of lympho-hematopoietic stem cells. Nat Immunol 3:329–333
4. de Haan G, Nijhof W, Van Zant G (1997) Mouse strain-dependent changes in frequency and proliferation of hematopoietic stem cells during aging: correlation between lifespan and cycling activity. Blood 89:1543–1550

5. Harrison DE, Astle CM, Delaittre JA (1978) Loss of proliferative capacity in immunohemopoietic stem cells caused by serial transplantation rather than aging. J Exp Med 147:1526–1531
6. Ford CE, Hamerton JL, Barnes DW, Loutit JF (1956) Cytological identification of radiation-chimaeras. Nature 177:452–454
7. McCulloch EA, Till JE (1960) The radiation sensitivity of normal mouse bone marrow cells, determined by quantitative marrow transplantation into irradiated mice. Radiat Res 13:115–125
8. Purton LE, Scadden DT (2007) Limiting factors in murine hematopoietic stem cell assays. Cell Stem Cell 1:263–270
9. Harrison DE (1980) Competitive repopulation: a new assay for long-term stem cell functional capacity. Blood 55:77–81
10. Rosendaal M, Hodgson GS, Bradle TR (1979) Organization of haemopoietic stem cells: the generation-age hypothesis. Cell Tissue Kinet 12:17–29
11. Purton LE, Dworkin S, Olsen GH, Walkley CR, Fabb SA, Collins SJ, Chambon P (2006) RAR{gamma} is critical for maintaining a balance between hematopoietic stem cell self-renewal and differentiation. J Exp Med 203: 1283–1293
12. Ema H, Morita Y, Yamazaki S, Matsubara A, Seita J et al (2006) Adult mouse hematopoietic stem cells: purification and single-cell assays. Nat Protoc 1:2979–2987
13. Mauch P, Hellman S (1989) Loss of hematopoietic stem cell self-renewal after bone marrow transplantation. Blood 74: 872–875
14. Yu H, Yuan Y, Shen H, Cheng T (2006) Hematopoietic stem cell exhaustion impacted by p18 INK4C and p21 Cip1/Waf1 in opposite manners. Blood 107:1200–1206
15. Kiel MJ, Yilmaz OH, Iwashita T, Yilmaz OH, Terhorst C, Morrison SJ (2005) SLAM family receptors distinguish hematopoietic stem and progenitor cells and reveal endothelial niches for stem cells. Cell 121:1109–1121

Chapter 4

Isolation, Characterization, and Transplantation of Adult Liver Progenitor Cells

Mladen I. Yovchev, Mariana D. Dabeva, and Michael Oertel

Abstract

Many chronic liver diseases are life-threatening. When the liver loses the ability to repair itself the only treatment currently available is liver transplant. However, there are not enough donors to treat all the patients. This requires the search of alternative therapies utilizing stem and progenitor cells for treatment of these patients and restoration of their normal liver function.

Hepatic progenitor cells can be isolated from livers at different developmental stages including adult liver. In the adult rat liver, there is clear evidence that progenitor cells (also called "oval cells") derive from precursors in the canals of Herring that are capable to differentiate into hepatocytes and bile duct cells. In experimental models, hepatic progenitor cells can be isolated and propagated in vitro and used for restoration of the diseased liver. The first step in utilization of progenitor cells is their identification in the liver, isolation of purified progenitor cell fractions, which are subsequently transplanted in the diseased liver for evaluation of liver repopulation by transplanted cells, and evaluation their potentials for clinical application.

The present protocol describes the isolation of non-parenchymal cells (NPCs) from wt DPPIV$^+$ F344 rats, followed by purification of "oval cells", immunohistochemical staining techniques to characterize these cells, their transplantation into retrorsine-treated mutant DPPIV$^-$ rats, as well as the enzyme histochemical staining for DPPIV to detect transplanted cells in the host liver.

Keywords Liver progenitor cells, "Oval cells", Non-parenchymal cells (NPCs), Progenitor cell isolation, Transplantation, Liver repopulation

1 Introduction

The existence of hepatic progenitor cells was postulated by Wilson and Leduc who observed proliferation of cholangiocytes that give rise to hepatocytes and new interlobular bile ducts in the liver of mice subjected to nutritional injury (1). These cells are small, with a pale, oval-shaped nucleus and scant cytoplasm and thus named "oval cells" by Farber (2). In the normal adult liver, progenitor cells are rare, however, "oval cell" activation can be induced by toxic agents (2), e.g., 2-acetylaminofluorene (2-AAF) in combination with 2/3 partial hepatectomy (PH) as regenerative stimulus (2),

Kursad Turksen (ed.), *Stem Cells and Aging: Methods and Protocols*, Methods in Molecular Biology, vol. 976, DOI 10.1007/978-1-62703-317-6_4, © Springer Science+Business Media, LLC 2013

D-galactosamine (3), or a choline-deficient diet (4). These cells express hepatocytic markers (e.g., α-fetoprotein, albumin, CK-8) biliary epithelial cell markers (e.g., CK-19, OV-6) (5), EpCAM, claudin-7, CD44 (6), and mesenchymal markers (e.g., vimentin, mesothelin, BMP-7) (6) (see Fig. 1).

Progenitor cell-enriched non-parenchymal cell (NPC) fractions can be isolated from enzymatically digested liver, by density gradient centrifugation, which separates different cell populations according to their density (7–9). Oval cell enriched NPC fractions contain up to 20% "oval cells". Additional purification steps need to be performed to achieve high cell enrichment levels which are necessary to further characterize and describe the properties and behavior of progenitor cells using RT-PCR or immunohistochemical staining techniques. A powerful tool to isolate highly purified progenitor cell populations (90–95%) from rat livers is the combination of density gradient centrifugation and magnetic cell sorting (MACS) technology using antibodies specific for known epitops on progenitor cells (9–12). However, most of the surface antigens expressed on "oval cells" are also expressed on biliary epithelial cells. Therefore, using a surface antigen to purify progenitor cells, high cell yield of "oval cells" can be achieved but the cell fractions will contain also bile duct cells (9) (see Fig. 2a, b).

To detect transplanted hepatic progenitor cells in the host liver and determine their proliferative capacity and differentiation potential into hepatic cell lineages, a cell transplantation model capable to follow the fate of infused cells has to be used. The normal Fischer (F)344 rat, which expresses the exopeptidase DPP4 on the surface of hepatocyte, biliary epithelial cells and progenitor cells can serve as the source for isolation of wild type (wt) $DPPIV^+$ donor cells, which can be transplanted into mutant $DPPIV^-$ recipients (13) without rejection (8, 14). The most sensitive technique to detect $DPPIV^+$ transplanted cells in the $DPPIV^-$ recipient is an enzyme histochemical staining procedure, using glycyl-proline-4-methoxy-β-naphthylamide as substrate (15). Importantly, to achieve efficient liver repopulation, the recipient liver has to be manipulated to provide an environment in which transplanted cells have a selective advantage over host hepatocytes (reviewed in ref. 16). This can be induced by retrorsine, a pyrrolizidine alkaloid that blocks DNA proliferation of mature hepatocytes. Therefore, retrorsine-preconditioned recipient liver cells cannot respond to a liver proliferative stimulus (e.g., partial hepatectomy) and transplanted hepatocytes (17) or "oval cells" (9) effectively replace the recipient liver (see Fig. 2c, d).

Schematic overview of procedures Liver tissues after 2-AAF administration or cytospins of isolated cell fractions can be immunohistochemical stained (IHC) for progenitor cell-specific markers to characterize and describe the properties of these cells in the adult liver (**IV**). Finally, isolated (and purified) progenitor cells from $DPPIV^+$ F344 rat livers are transplanted into the liver of retrorsine (RS)-preconditioned $DPPIV^-$ mutant F344 rats immediately after 2/3 PH and repopulation is followed over time by enzyme histochemistry (EHC) for DPPIV (**V**)

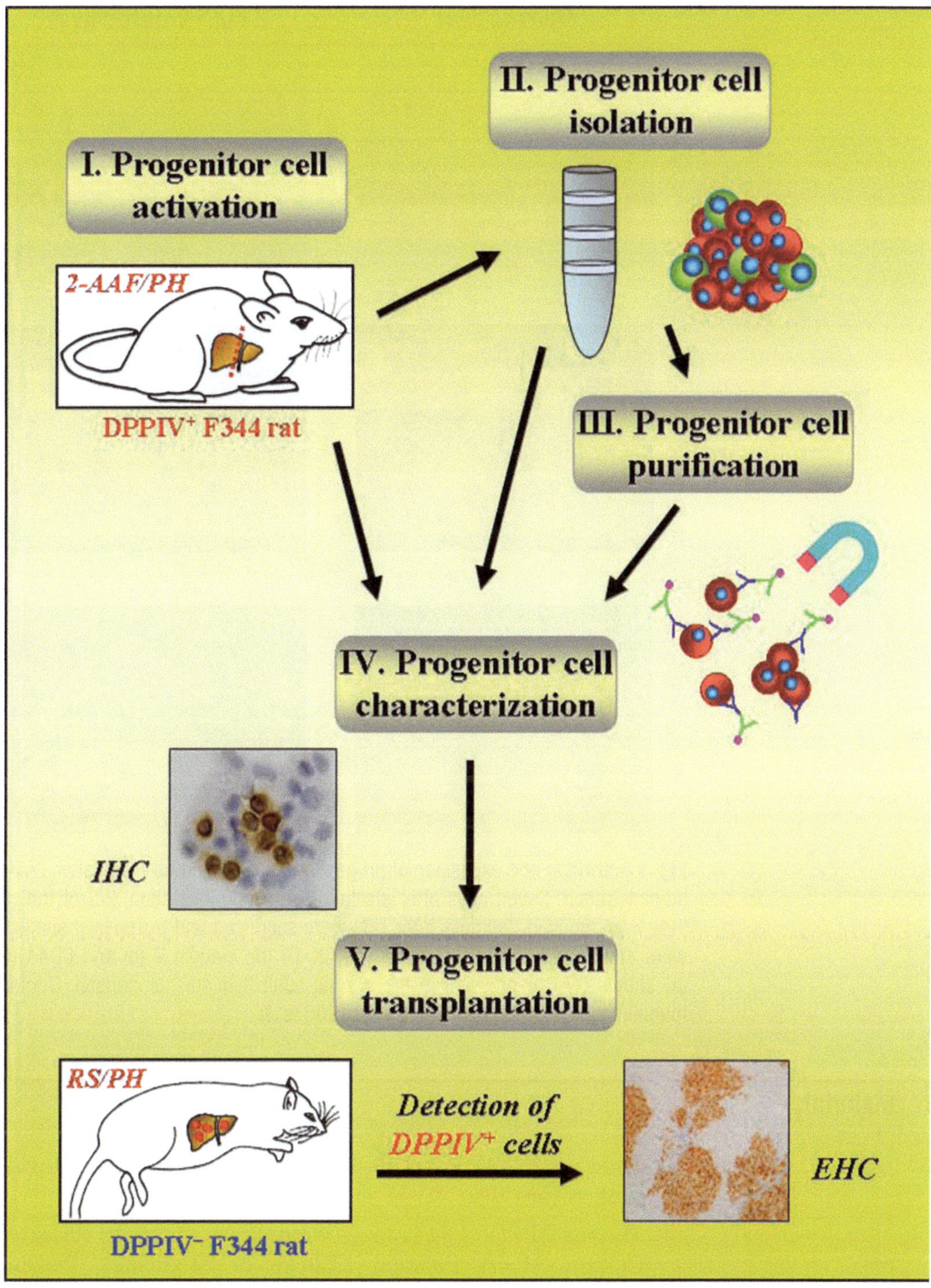

Schematic overview of procedures described in the protocol of this chapter. Activation of progenitor cells in the adult liver of wt (dipeptidyl peptidase (DPP)IV+) F344 rats by 2-acetyl aminofluorene (2-AAF) administration in combination with 2/3 partial hepatectomy (PH) (**I**) Non-parenchymal cells (NPCs) are isolated through enzymatic digestion of the liver, followed by density gradient centrifugation to separate cell populations (**II**). Progenitor cells are further purified by magnetic microbeads cell sorting using an antibody specific for EpCAM (**III**).

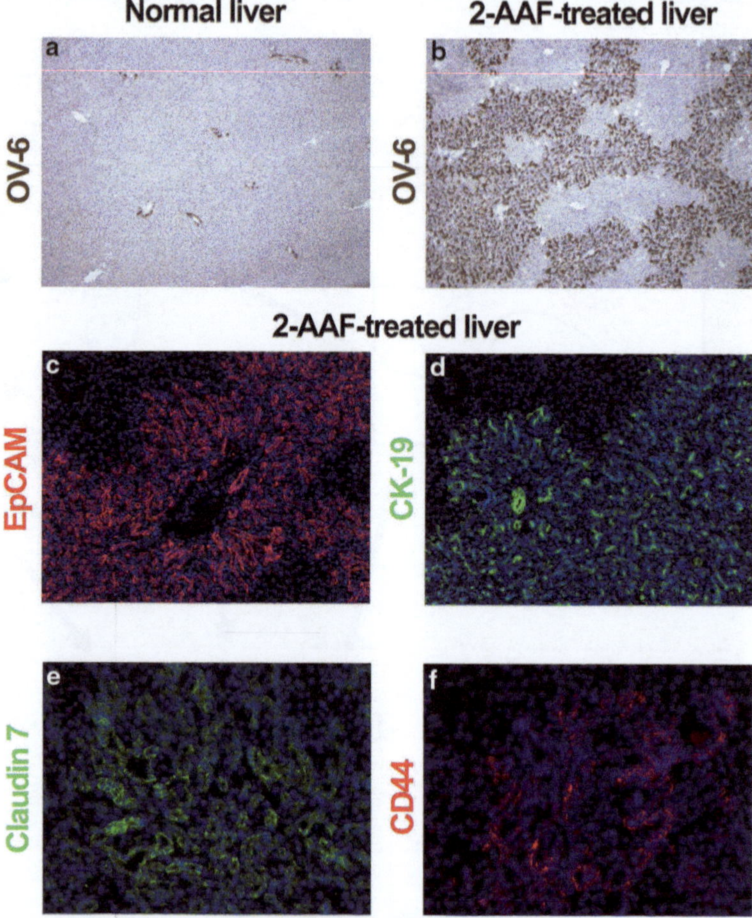

Fig. 1 Activation and expansion of progenitor cells in adult rat livers after 2-AAF administration. Seven days after starting 2-AAF administration, 2/3 of the rat liver was removed. Ten days later, rats were sacrificed and frozen liver sections were stained for OV-6 (**b**), EpCAM (**c**), CK-19 (**d**), claudin 7 (**e**) and CD44 (**f**). (**a**) shows OV-6 expression in the normal adult liver for comparison. Original magnification, ×50 (**a**, **b**), ×100 (**c**, **d**), ×200 (**e**, **f**)

2 Materials

2.1 Animals

1. Male wt F344 rats, 180–200 g (Taconic Farms). Provide progenitor cells.
2. Mutant DPPIV$^-$ F344 rats, 2–4 months of age (Liver Research Center, Albert Einstein College of Medicine). Provide recipient rats.

2.2 Chemicals and Reagents

1. 2-acetylaminofluorene (2-AAF), 35 mg pellets with 14-days time release of 2-AAF (Innovative Research of America).
2. Retrorsine (Sigma).

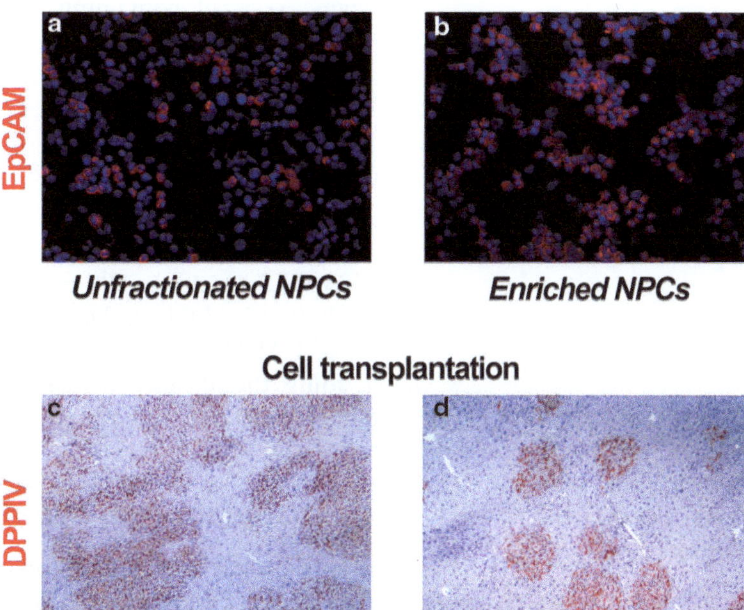

Fig. 2 Enrichment of adult progenitor cells and their transplantation into preconditioned rat livers. (**a**, **b**) Magnetic bead cell sorting was used to enrich EpCAM+ progenitor cells. Isolated NPCs contain up to 20% EpCAM+ cells (**a**) and more than 90% after microbeads-enrichment (**b**). (**c**, **d**) Detection of differentiated DPPIV+ cells derived from transplanted progenitor cells in retrorsine-treated rats. One month (**c**) and 14 days (**d**) after transplantation of 4×10^7 NPCs (**c**) and 2.5×10^6 purified EpCAM+ cells (**d**) into mutant DPPIV− F344 rats in conjunction with 2/3 partial hepatectomy, DPPIV+ hepatocytic clusters were detected by enzymehistochemistry in frozen liver tissues. Original magnification, ×400 (**a**, **b**), ×40 (**c**, **d**)

3. DAB (Diaminobenzidine) (Sigma).
4. DAPI (4′,6-diamidino-2-phenylindole dichloride) can be purchased from many companies.
5. Gly-Pro-4-Methoxy-Beta-Naphtylamide (GPMN) (Polysciences).
6. Dimethyl-formamide (DMF) (Sigma).
7. Fast blue BB salt (Sigma).

2.3 Solutions

2.3.1 Solutions for NPC Isolation

1. *NPC dissociation buffer* (*buffer I*): To prepare 500 mL of Hank's balanced salt solution (HBSS) containing 2 mM $CaCl_2$ and 20 mM HEPES (pH 7.4), dilute 50 mL of a 10× HBSS solution (without Ca^{2+}; phenol red) (GIBCO) by adding 400 mL ddH_2O. Add 1 mL of 1 M $CaCl_2$ and 10 mL of 1 M HEPES (pH 7.4) stock solutions. Using a pH meter, adjust the pH to 7.4 with 1 M NaOH. Bring to a final volume of 500 mL with ddH_2O.

2. *NPC digestion enzyme cocktail* (see Note 1): To prepare 100 mL of enzyme cocktail containing 0.1% collagenase (Type I), 0.02% pronase E and 0.004% DNAse I (*immediately before procedure*), dissolve 100 mg collagenase (Worthington Biochemical) and 20 mg pronase E (EMD Chemicals) in 100 mL *buffer I.* Add 400 μL of a 1% DNAse I solution (Boehringer Mannheim).
3. *NPC washing buffer* (*buffer II*): To prepare 1,000 mL of HBSS containing 0.2% bovine serum albumin (BSA) and 20 mM HEPES (pH 7.4), dilute 100 mL of a 10× HBSS solution by adding 800 mL ddH_2O. Dissolve 2 g BSA (SIGMA) and 20 mL of 1 M HEPES (pH 7.4) stock solution. Using a pH meter, adjust the pH to 7.4 with 1 M NaOH. Bring to a final volume of 1,000 mL with additional ddH_2O.

2.3.2 Solutions for Progenitor Cell Fractionation

OptiPrep solutions: The following density gradient solutions have to be prepared (*immediately before procedure*):

11% OptiPrep solution: mix 14.6 mL of a 60% OptiPrep stock solution (Axis-Shield) and 65.4 mL *buffer II.*

13% OptiPrep solution: mix 17.3 mL of a 60% OptiPrep stock solution and 62.7 mL *buffer II.*

16% OptiPrep solution: mix 21.3 mL of a 60% OptiPrep stock solution and 58.7 mL *buffer II.*

18% OptiPrep solution: mix 12 mL of a 60% OptiPrep stock solution and 28 mL *buffer II.*

2.3.3 Solutions for Progenitor Cell Purification

1. *Cell sorting buffer*: To prepare 500 mL of HBSS containing 0.3% BSA, 0.8 mM $MgCl_2$ and 10 mM HEPES (pH 7.4), dilute 50 mL of a 10× HBSS solution by adding 440 mL ddH_2O and dissolve 1.5 g BSA. Add 400 μL of 1 M $MgCl_2$ and 5 mL of 1 M HEPES (pH 7.4) stock solutions. Add 5 mL of a penicillin/streptomycin solution (final concentrations of 100 U/mL and 100 μg/mL, respectively).

2.3.4 Chromogen (DAB) Visualization Solution

DAB solution: Add 3 mg 3,3′-Diaminobenzidine (DAB) (SIGMA) to 5 mL of a 50 mM Tris–HCl/10 mM immidazole solution (pH 7.4). Immediately before use, add 10 μL of a 30% peroxide solution (final concentration of 0.06%) and mix well.

2.3.5 Solutions for Histochemical Detection of DPP4

1. *TMS buffer* (0.1 M Trisma® maleate/0.1 M NaCl; pH 6.5):

   Step I: To prepare 200 mL of 1 M Trisma® maleate stock solution, dissolve 47.4 g Trisma® maleate in 140 mL ddH_2O and adjust the pH to 6.5 with 10N NaOH. Bring to a final volume of 200 mL with ddH_2O.

   Step II: To prepare 500 mL of 0.1 M Trisma® maleate/0.1 M NaCl (TMS) buffer solution, add 10 mL of a 5 M NaCl stock solution to 50 mL of 1 M Trisma® maleate stock solution. Bring to a final volume of 500 mL with ddH_2O.

2. *Substrate solution*: Dissolve 100 mg Fast blue BB salt (SIGMA) in 100 mL TMS buffer. Dissolve 50 mg glycyl-proline-4-methoxy-β-naphtylamide (GPMN) (SIGMA) in 3 mL DMF. Mix both solutions before use.
3. *Scott's water*: Dissolve 100 mg $MgSO_4 \times 7H_2O$ and 10 g $NaHCO_3$ in 500 mL ddH_2O (10× stock solution). Dilute 1:10 in ddH_2O before use.

2.4 Antibodies

1. *Primary antibodies*: mouse anti-OV-6 antibody (R&D Systems), mouse anti-EpCAM antibody (BioVendor), mouse anti-CK-19 antibody (Novocastra), rabbit anti-Claudin 7 antibody (Abcam), mouse anti-CD44 antibody (AbD Serotec).
2. *Secondary antibodies*: HRP-conjugated horse anti-mouse IgG (GE Healthcare), Cy™3-conjugated donkey anti-mouse IgG, Cy™2-conjugated donkey anti-mouse IgG, Cy™2-conjugated donkey anti-rabbit IgG (Jackson Immunoresearch).
3. *Mounting media*:
   (a) Vectashield mounting medium (Vector Lab.)
   (b) Permount (Fisher Scientific)

3 Methods

3.1 Progenitor Cell Activation

3.1.1 2-AAF Administration

1. Anesthetize the rat with ether or 1–5% isoflurane inhalation.
2. Shave the right lateral side of the neck between the ear and shoulder and disinfect the skin with 70% ethanol and let it dry.
3. Using dry sterile instrument, make a small (~1 cm) incision through the skin.
4. Spread the subcutaneous tissue to create enough space for the pellet and insert the 2-AAF pellet between skin and muscle.
5. Close the wound with suture and clean with sterile saline.

3.1.2 Partial Hepatectomy

1. Seven days after implantation of the 2-AAF pellet, anesthetize the rat, shave and clean the abdomen.
2. Make a 3–5 cm long midline incision through the skin and abdominal wall up to the xiphoid process.
3. Lift the xiphoid process and cut the falciform ligament which connects the under surface of the diaphragm and the liver.
4. Gently press the abdomen at both sides of the incision with forefingers and thumbs and push until the median and left lateral lobes are exposed.

5. Using 2-0 silk suture, ligate the liver at the base of the exposed liver lobes. Lift and cut the lobes above to the ligature, using curved scissors.
6. After finishing partial hepatectomy, add 1 mL of sterile saline containing 100 U penicillin and 100 μg streptomycin into the peritoneal cavity. Close the peritoneal wall with absorbable suture and use silk suture for the dermal layers of the skin. Clean the wound with sterile saline, using cotton-tipped applicators.
7. After 10 days, proceed with liver perfusion and progenitor cell isolation.

3.2 Progenitor Cell Isolation and Fractionation

Before starting with the isolation of NPCs, the adult rat liver has to be perfused using a two-step digestion method. This standard procedure was described in detail by D. Neufeld (18).

1. Dissect and place the perfused liver in a sterile 10-cm Petri dish containing 15 mL chilled RPMI with 10% fetal bovine serum (FBS) (Biowest).
2. Disintegrate the liver tissue using a cell scraper (see Note 2). Add an equal volume of cold *buffer II* to the Petri dish. Collect cell suspension.
3. Repeat step 2 several times until the liver is almost completely disintegrated.
4. Filter the collected cell suspensions through an 80-μm nylon mesh placed over a 200-mL glass beaker and proceed with step 5. Collect the remaining undigested liver fragments in 10 mL of *enzyme cocktail* and the tissue fragments on ice for later use (see step 9).
5. Dilute cell suspension with *buffer II* and bring to a final volume of 320 mL. Divide cell suspension into eight 50-mL centrifuge tubes, 40 mL each.
6. Centrifuge at 50 × *g* for 30 s at 25°C.
7. Collect the supernatants containing NPCs and divide cell suspension into eight new 50-mL centrifuge tubes. Discard cell pellet.
8. Centrifuge at 400 × *g* for 10 min at 25°C.
9. Discard all supernatants. Resuspend remaining cell pellets in 25 mL of *enzyme cocktail* and transfer cell suspension into a 500-mL glass flask containing 15–20 glass beads. Add the undigested liver fragments, which were kept in 10 mL *enzyme cocktail* on ice (see step 4), and add 65 mL of the *enzyme cocktail*.
10. Place the glass flask in a water bath and incubate for 30 min at 37°C under gentle agitation. Monitor the digestion process by checking the cell viability by trypan blue exclusion and digestion rate every 10 min.

11. Stop the reaction with 10 mL heat-inactivated FBS (final concentration 10% FBS) and filter cell suspension through a 40-μm nylon mesh placed over a 200-mL glass beaker.
12. Bring cell suspension to a total volume of 160 mL with *buffer II* and divide cell suspension into four 50-mL centrifuge tubes, 40 mL each.
13. Centrifuge at 50 × *g* for 1 min at 4°C. Discard cell pellet.
14. Collect the supernatants containing NPCs and divide cell suspension into four new 50-mL centrifuge tubes. Centrifuge at 400 × *g* for 10 min at 4°C.
15. Discard all supernatants. Wash cell pellets three times in *buffer II* (400 × *g*, 10 min, 4°C).
16. Resuspend all cell pellets in 80 mL of 11% OptiPrep solution and filter through a 40-μm cell strainer (see Note 3).
17. Divide 60 mL of the cell suspension into six polyallomer centrifuge tubes (25 × 89 mm) (Beckman), 10 mL each. The remaining 20 mL cell suspension will be used in step 20.
18. Using a 10-mL serological pipette, carefully underlay 10 mL of 13% OptiPrep solution in each tube.
19. Subsequently, underlay 10 mL of 16% OptiPrep solution and than 5 mL of 18% OptiPrep solution in each tube. A clear well defined border between the OptiPrep solutions will be seen.
20. Use the remaining cell suspension from step 18 to fill the tubes and balance the weight of all polyallomer centrifuge tubes. Place the tubes into the rotor buckets of a SW-28 swinging rotor (Beckmann).
21. Centrifuge at 6,500 × *g* for 30 min at 4°C without brake.
22. After centrifugation, carefully remove the tubes from the rotor buckets and place them into a tube holder. Three cell layers will be clearly visible: between 11 and 13% (NPC fraction 1), between 13 and 16% (NPC fraction 2), and between 16 and 18% (NPC fraction 3) Optiprep (see Note 4).
23. Using a 10-mL serological pipette, carefully collect cells located between the 11 and 13% OptiPrep layers (NPC fraction 1) and transfer them into a 50-mL centrifuge tube. Similarly, collect cells located between the 13 and 16% OptiPrep layers (NPC fraction 2). Bring to a final volume of 45 mL with *buffer II* in both tubes. If further purification of oval cells is not needed collect only the cells at 13–16% boundary (NPC fraction 2).
24. Centrifuge at 800 × *g* for 10 min at 4°C.
25. Combine cell pellets from both NPC fractions 1 and 2 and wash cells two times in 45 mL *buffer II* and centrifuge at 400 × *g* for 5 min at 4°C (see Note 5).

3.3 Progenitor Cell Purification

1. Dilute isolated NPCs (see Subheading 3.2; Note 6) in cell sorting buffer and bring to a final volume of 45 mL.
2. Centrifuge at 300 × *g* for 5 min at 4°C.
3. Resuspend the pellet in cell sorting buffer and adjust 1×10^8 cells per 2 mL buffer.
4. Add the monoclonal mouse anti-EpCAM (BioVendor). The recommended concentration is 0.1–0.15 μg/10^6 total cells (see Notes 7 and 8).
5. Incubate cell suspension containing primary antibody for 30 min at 4°C under gentle agitation.
6. Wash cells in 45 mL cell sorting buffer and centrifuge at 300 × *g* for 5 min at 4°C. Repeat this step once.
7. Resuspend cell pellet in cell sorting buffer and adjust 1×10^7 cells per 160 μL buffer. Add secondary goat anti-mouse IgG1 antibody bound to magnetic microbeads (40 μL per 160 μL cell suspension) (Miltenyi Biotec).
8. Incubate cell suspension containing microbeads-conjugated antibody for 30 min at 4°C under gentle agitation.
9. Repeat step 6.
10. Resuspend cell pellet in cell sorting buffer and adjust 5×10^7 cells per 5 mL buffer.
11. Place a LS column (Miltenyi) assembled with a 30-μm Pre-separation filter (Miltenyi) in a MidiMACS™ Separator (Miltenyi) and rinse with 2 mL cell sorting buffer.
12. Slowly apply cell suspension onto Pre-separation filter and LS column (see Note 9).
13. Wash LS column with 2 mL cell sorting buffer.
14. Carefully remove LS column from the magnet and place it into a 15-mL centrifuge tube. Rinse LS column with 5 mL cell sorting buffer. Gently use the column-plunger to collect the positive-selected cells.
15. Repeat steps 11–14, using a new LS column.
16. Wash cell suspension two times in 45 mL cell sorting buffer.

3.4 Progenitor Cell Identification

3.4.1 Tissue Preservation/Fixation

After sacrifice of the rat, remove the liver and cut liver lobes into several pieces. To avoid crystal formation, dab the liver tissue sample with a paper towel to remove excessive liquid. Add Tissue-Tek® OCT Compound (Fisher Scientific) on a piece of cork (~1 × 1 in.), press specimen flat into the OCT and drop the sample immediately into chilled 2-methylbutane (prechilled in a beaker on dry-ice). Let liver tissue samples freeze for at least 5 min and store at −80°C until use. (Alternatively, you can use Tissue-Tek cryomolds Miles, Inc.)

3.4.2 Identification of Progenitor Cells in Tissue Sections. Immunohistochemical Detection

1. Cut fresh sections of 5 μm from snap-frozen block of liver tissue and transfer on *Colorfrost®/Plus* microscope slides (Fisher Scientific).
2. Air-dry the cryosections for 5 min at RT.
3. Fix cryosections for 10 min in ice-cold 100% methanol.
4. Air-dry the slides for 5 min at RT.
5. Encircle the liver sections with PAP PEN (Polysciences) and dry for 2 min at RT.
6. Rehydrate sections for 20 min in PBS.
7. Block endogenous peroxidase by immersing liver sections in 3% H_2O_2 solution in methanol for 30 min at RT. Wash slides with PBS for 20 min.
8. To block nonspecific binding of secondary antibody, incubate liver sections in PBS (containing 2% horse (Vector Laboratories) or donkey serum (SIGMA), 2% BSA and 0.05% Tween 20) for 1 h at RT in a humidified chamber (see Note 10).
9. Dilute primary antibodies to its optimal concentration in PBS (containing 2% horse or donkey serum, 2% BSA and 0.05% Tween 20), apply antibody to the sections and incubate for 2 h at RT *or* overnight at 4°C in a humidified chamber (see Note 11).
10. Wash the slides three times for 10 min with PBS.
11. Dilute secondary, HRP-conjugated antibodies in PBS (containing 2% rat serum (SIGMA), 2% BSA and 0.05% Tween 20) at 1:100 and incubate tissue sections for 60 min at RT in a humidified chamber.
12. Wash the slides three times for 10 min with PBS.
13. Cover liver sections with DAB solution and incubate for 10–20 min until good staining intensity is achieved.
14. Wash slides with PBS for 5 min.
15. Counter stain liver sections with Harris hematoxylin for 3–5 min. Rinse in water for 5 min.
16. To dehydrate tissue sections, place slides in a staining jar with 50% ethanol for 3 min. Transfer slides to 75%, 95% and then to 100% ethanol for 3 min each step.
17. Soak slides two times for 3 min in xylene.
18. Mount tissue sections with Permount® mounting medium and cover with glass coverslip.

3.4.3 Identification of Progenitor Cells in Tissue Sections. Immunofluorescent Detection

1. Perform steps 1–6.
2. Omit step 7.
3. Continue with steps 8 and 9. All procedures after this step should be done in the dark.

4. Dilute secondary antibody in PBS (containing 2% rat serum (SIGMA), 2% BSA and 0.05% Tween 20) and incubate tissue sections for 60 min at RT in a humidified chamber. (Different secondary antibodies conjugated to different fluorescent dyes (Cy[Tm]2, Cy[Tm]3, FITC, PE, TRITC, etc.) are commercially available. The optimal dilution should be determined by the user.)
5. Wash the slides three times for 10 min with PBS.
6. Incubate the sections with 0.5 μg/mL of DAPI (4′,6-diamidino-2-phenylindole dichloride), diluted in PBS to visualize cell nuclei for 3–5 min.
7. Wash the slides 5 min in PBS.
8. Mount tissue sections with Vectashield and cover with glass coverslip.
9. Examine liver sections using a fluorescence microscope with appropriate filter sets.

3.4.4 Identification of Progenitor Cells in Cytospins

The described immunohistochemical staining and immunofluorescent procedures can be used also for cytospins. To prepare cytospins from isolated cell fractions, centrifuge ~1–1.5 × 10^4 cells/300 μL medium per slide at 90 × *g* for 5 min at RT in a cytospin centrifuge.

3.5 Progenitor Cell Transplantation

3.5.1 Preconditioning of the Recipient Animals with Retrorsine Before Cell Transplantation

1. Dissolve 30 mg retrorsine (SIGMA) in 1.5 mL 0.5 M HCl (see Note 12).
2. Add dropwise 0.5 M NaOH and adjust the pH to 7.0, using pH indicator paper.
3. Bring to a final volume of 5 mL with additional sterile ddH_2O.
4. Inject retrorsine solution (30 mg/kg b.w.; i.p.) into mutant DPPIV$^-$ F344 rats weighing 100–150 g, using a 26 G½ gauge needle connected to a 1-mL syringe. Repeat retrorsine injection after 2 weeks.
5. After additional 4 weeks, preconditioned rats can be used as recipients for cell transplantation studies.

3.5.2 Preparation of the Cells for Transplantation

1. Dilute isolated NPCs (see Subheading 3.2) or purified "oval cells" (see Subheading 3.3) and wash the cell suspension once in 45 mL Dulbecco's modified Eagle medium (DMEM) (GIBCO), containing 10% heat-inactivated FBS, and centrifuge at 400 × *g* for 5 min at 4°C.
2. Resuspend the cell pellet in 1 mL DMEM with 10% FBS and filter cell suspension through a 40-μm cell strainer.
3. Determine cell number and count for viable cells by trypan blue exclusion. After adjusting appropriate cell number, keep cell suspension on ice until cell transplantation.

3.5.3 Cell Infusion

1. Proceed with steps 1–5 as described in Subheading 3.1.2.
2. After finishing partial hepatectomy, carefully expose the spleen using forceps and cotton-tipped applicators.
3. The inferior pole of the spleen has to be loosely ligated prior to cell infusion, using 2-0 silk suture.
4. Inject 0.5–1.0 mL cell suspension through the inferior side of the spleen, using a 26 G½ gauge needle connected to a 1-mL syringe. After removing the needle, quickly tighten the ligation to prevent cell leakage or bleeding after cell injection.
5. Add few drops of sterile saline containing antibiotics into the peritoneal cavity and close the peritoneal wall using absorbable suture. Suture the abdominal skin and clean the wound with saline.
6. Finally, inject sterile saline subcutaneously into the dorsal surface of neck to support recovery of the rat.

3.5.4 Enzymatic Histochemical Detection of DPPIV

Use 5 μm cryosections (see Subheading 3.4, step 1).

1. After drying, fix liver sections in prechilled 95% ethanol–glacial acetic acid (99:1 vol/vol) for 5 min, followed by a wash step in cold 95% ethanol for 5 min.
2. Air-dry completely the slides at RT.
3. Incubate slides in the substrate solution (Koplik jar) for 30–45 min at 37°C (see Note 13).
4. Wash liver sections three times in TMS buffer at RT, 2 min each step (see Note 14).
5. Incubate slides in 0.1 M CuSO4 solution for 5 min to stabilize red color.
6. Rinse liver sections three times in TMS buffer, 2 min each step.
7. Fix liver sections in 4% PFA solution (made in TMS or saline) for 10 min at 4°C.
8. Wash three times for 10 min in TMS buffer at RT and rinse briefly with water.
9. Counterstain liver sections with Harris hematoxylin for 1–3 min, followed by a wash step in water for 5 min.
10. For a crisp blue-violet nuclear counterstaining, incubate the slides for 20 s in Scott's water and rinse with water, thereafter.
11. Air-dry liver sections for 30 min. Slides can be permanently stored in the dark at RT.
12. For evaluation under a microscope, mount the slides in pure glycerol and cover with a glass coverslip. After evaluation, remove the glass coverslip, rinse slides with water, and air-dry to preserve the staining.

4 Notes

1. The quality of enzymes is crucial for the isolation procedure. Different enzyme preparations (even from the same vendor) might have different properties and activity. Therefore, before purchasing commercially available collagenase or pronase, test a sample of a particular lot by performing a cell isolation and check cell yield and viability, thereafter.
2. If the liver was well perfused, hepatic cells will be easily released and only few undigested liver pieces will remain.
3. Cells must be well resuspended in an appropriate volume to allow distribution of NPCs between the layers.
4. NPC fraction 1 is enriched in stellate cells, blood cells and fibroblasts but it also contains oval cells. Cells collected between the 16 and 18% OptiPrep layers (NPC fraction 3) have to be discarded, because this fraction is usually contaminated with hepatocytes.
5. Using this protocol, nearly 1.5×10^8 total NPCs can be isolated. However, the number of "oval cells" differs from animal to animal and depends mainly on the progenitor cell activation induced through 2-AAF administration.
6. Before starting the procedure, make sure that pronase was omitted from the liver digestion enzyme cocktail during NPCs isolation. This enzyme destroys the extracellular domain of the surface proteins. Keep this in mind also for immunohistochemical staining procedures.
7. Optimal antibody dilution should be determined by the investigator.
8. Although several antibodies detecting "oval cells" have been reported, only few of them can be successfully used for cell separation. The ideal "oval cells"-specific antibody should be highly specific and must recognize epitops on the extracellular domain of the cell surface molecules. In our hands, the best antibody to purify "oval cells" was monoclonal mouse anti-EpCAM (clone GZ1 and clone GZ20).
9. $EpCAM^+$ cells will be retained in the magnetic field of the LS column; $EpCAM^-$ cells will pass through the LS column.
10. Depending on the source of secondary antibody, use heat-inactivated (56°C for 30 min) serum from the same species.
11. In our hands, optimal dilutions of primary antibodies against OV-6, EpCAM, CK-19, Claudin 7 and CD44 were 1:100, 1:50, 1:100, 1:50 and 1:100, respectively. Nevertheless, check the optimal working dilutions before use.
12. Retrorsine needs to be dissolved completely before neutralization step.

13. If few slides need to be processed, the substrate solution can be scaled down and the solution pipette over the section.
14. 0.14 M NaCl solution can be used instead of TMS buffer for all wash steps throughout the procedure.

References

1. Wilson JW, Leduc EH (1958) Roles of cholangioles in restoration of the liver of the mouse after dietary injury. J Pathol Bacteriol 76: 441–449
2. Farber E (1956) Similarities in the sequence of early histological changes induced in the liver of the rat by ethionine, 2-acetylamino-fluorene, and 3′-methyl-4-dimethylaminoazobenzene. Cancer Res 16:142–148
3. Lemire JM, Shiojiri N, Fausto N (1991) Oval cell proliferation and the origin of small hepatocytes in liver injury induced by D-galactosamine. Am J Pathol 139:535–552
4. Sells MA, Katyal SL, Shinozuka H et al (1981) Isolation of oval cells and transitional cells from the livers of rats fed the carcinogen DL-ethionine. J Natl Cancer Inst 66:355–362
5. Oertel M, Shafritz DA (2008) Stem cells, cell transplantation and liver repopulation. Biochim Biophys Acta 1782:61–74
6. Yovchev MI, Grozdanov PN, Joseph B et al (2007) Novel hepatic progenitor cell surface markers in the adult rat liver. Hepatology 45: 139–149
7. Yaswen P, Nancy T, Hayner NT, Fausto N (1984) Isolation of oval cells by centrifugal elutriation and comparison with other cell types purified from normal and preneoplastic livers. Cancer Res 44:324–331
8. Dabeva MD, Hwang S-G, Vasa SRG et al (1997) Differentiation of pancreatic epithelial progenitor cells into hepatocytes following transplantation into rat liver. Proc Natl Acad Sci USA 94:7356–7361
9. Yovchev MI, Grozdanov PN, Zhou H et al (2008) Identification of adult hepatic progenitor cells capable of repopulating injured rat liver. Hepatology 47:636–647
10. Yovchev MI, Zhang J, Neufeld DS et al (2009) Thymus cell antigen-1-expressing cells in the oval cell compartment. Hepatology 50:601–611
11. Oertel M, Menthena A, Chen Y-Q et al (2007) Comparison of hepatic properties and transplantation of Thy1^{+} and Thy-1^{-} cells isolated from ED14 rat fetal liver. Hepatology 46: 1236–1245
12. Oertel M, Menthena A, Chen Y-Q et al (2008) Purification of fetal liver stem/progenitor cells containing all the repopulation potential for normal adult rat liver. Gastroenterology 134: 823–832
13. Thompson NL, Hixson DC, Callanan H et al (1991) A Fischer rat substrain deficient in dipeptidyl peptidase IV activity makes normal steady-state RNA levels and an altered protein. Use as a liver-cell transplantation model. Biochem J 273:497–502
14. Rajvanshi PA, Kerr A, Bhargava KK et al (1996) Studies of liver repopulation using the dipeptidyl peptidase IV deficient rat and other rodent recipients: cell size and structure relationships regulate capacity for increased transplanted hepatocytes mass in the liver lobule. Hepatology 23:482–496
15. Lojda Z (1979) Studies on dipeptidyl(amino) peptidase IV (glycyl-proline naphthylamidase). II. Blood vessels. Histochemistry 59:153–166
16. Oertel M (2011) Fetal liver cell transplantation as a potential alternative to whole liver transplantation? J Gastroenterol 46:953–965
17. Laconi E, Oren R, Mukhopadhay D et al (1998) Long-term, near total liver replacement by transplantation of isolated hepatocytes. Am J Pathol 153:319–329
18. Neufeld DS (1997) Isolation of rat liver hepatocytes. Methods Mol Biol 75:145–151

Chapter 5

Isolation of Muscle-Derived Stem/Progenitor Cells Based on Adhesion Characteristics to Collagen-Coated Surfaces

Mitra Lavasani*, Aiping Lu*, Seth D. Thompson, Paul D. Robbins, Johnny Huard, and Laura J. Niedernhofer

Abstract

Our lab developed and optimized a method, known as the modified pre-plate technique, to isolate stem/progenitor cells from skeletal muscle. This method separates different populations of myogenic cells based on their propensity to adhere to a collagen I-coated surface. Based on their surface markers and stem-like properties, including self-renewal, multi-lineage differentiation, and ability to promote tissue regeneration, the last cell fraction or slowest to adhere to the collagen-coated surface (pre-plate 6; pp6) appears to be early, quiescent progenitor cells termed muscle-derived stem/progenitor cells (MDSPCs). The cell fractions preceding pp6 (pp1–5) are likely populations of more committed (differentiated) cells, including fibroblast- and myoblast-like cells. This technique may be used to isolate MDSPCs from skeletal muscle of humans or mice regardless of age, sex or disease state, although the yield of MDSPCs varies with age and health. MDSPCs can be used for regeneration of a variety of tissues including bone, articular cartilage, skeletal and cardiac muscle, and nerve. MDSPCs are currently being tested in clinical trials for treatment of urinary incontinence and myocardial infarction. MDSPCs from young mice have also been demonstrated to extend life span and healthspan in mouse models of accelerated aging through an apparent paracrine/endocrine mechanism. Here we detail methods for isolation and characterization of MDSPCs.

Keywords Pre-plate technique, Muscle-derived stem/progenitor cells, Muscle, Adult stem cells, Regenerative medicine

1 Introduction

Adult stem cells hold great promise for regenerative medicine. Stem cells isolated from adult organisms have significant advantages over embryonic and fetal stem cells as therapeutic modalities. First, adult stem cells by-pass the substantial regulatory burden that accompanies the use and study of cells isolated from perinatal organisms (1–3). Second, adult stem cells offer the possibility of autologous therapy. Using cells isolated from a patient to treat only that patient eliminates the risk of an immune rejection that could

*Mitra Lavasani and Aiping Lu contributed equally.

Kursad Turksen (ed.), *Stem Cells and Aging: Methods and Protocols*, Methods in Molecular Biology, vol. 976,
DOI 10.1007/978-1-62703-317-6_5, © Springer Science+Business Media, LLC 2013

attenuate the therapeutic benefit, or worse cause significant side effects (4). In addition, autologous approaches for cell-based therapies will accelerate the path to the clinic (5–7).

Adult stem cells have been isolated from the bone marrow (8, 9), brain (10, 11), skeletal muscle (12–14), fat (15), skin (16), and gut (17–19). Muscle-derived stem cells offer several advantages over other adult stem cell populations for therapeutic applications. First, skeletal muscle is relatively easy to access, unlike, for example, brain and bone marrow. Second, a relatively small amount of tissue is necessary to isolate a sufficient number of stem cells for treatment purposes (4). Third, functional MDSPCs can be expanded substantially ex vivo (20). This minimizes the amount of tissue that must be excised from a patient, but also offers the possibility of treating a patient multiple times after a single procedure to isolate stem cells. Fourth, there is evidence that MDSPCs isolated from older organisms, which have lost their potency, can be rejuvenated to behave more like MDSPCs isolated from young adults (21). This means that even the elderly may be eligible for effective autologous therapy. Fifth, it is established that MDSPCs can be transduced with retroviral vectors expressing novel genes while maintaining their stem-like properties (22–30). MDSP-like cells isolated from humans show high regenerative potential in skeletal and cardiac muscle (31, 32). Thus, MDSPCs may potentially be used to treat inherited diseases of muscle degeneration such as muscular dystrophies. MDSPCs can differentiate into myogenic, osteogenic, chondrogenic, adipogenic, neural, endothelial, and hematopoietic cells (13, 33–45). Therefore, MDSPCs may be useful for treating degenerative diseases of multiple organ systems. Finally, MDSPCs appear to elicit a therapeutic benefit, at least in part, by secreting factors that promote host rejuvenation (21). Thus, therapeutic benefits may be achieved with fewer cells.

To date, MDSPCs have been used to improve bone healing (26, 35, 46), articular cartilage repair (36, 47), cardiac ischemia (48, 49), urinary incontinence (7, 38, 50–53), aging-related degenerative changes (21) and muscular injury or disease including muscular dystrophy (13, 34, 54). Although MDSPCs are not a well-defined, homogenous cell type, their tremendous potential for treating a wide spectrum of traumatic and degenerative changes supports the continued development and analysis of MDSPCs. Herein is described in detail the methods necessary for the isolation and minimal characterization of MDSPCs.

2 Materials

2.1 Reagents

1. Hank's Buffered Salt Solution *plus* Calcium Chloride and Magnesium Chloride (HBSS; Invitrogen, Cat. #24020-117, see Note 1).
2. Collagenase type XI (Sigma-Aldrich, Cat. #C7657).

3. Dispase (Invitrogen, Cat. #17105-041).
4. Trypsin-EDTA (Invitrogen, Cat. #15400-054).
5. Dulbecco's Modified Eagle Medium (DMEM, high glucose; Invitrogen, Cat. #11995-073).
6. Fetal Bovine Serum (FBS; Invitrogen, Cat. #10437-028).
7. Horse Serum (HS; Invitrogen, Cat. #26050-088).
8. Chick Embryo Extract (CEE; Accurate Chemical Co. Cat. #CE650T-10) (see Note 2).
9. Penicillin/Streptomycin (P/S, Invitrogen, Cat. #15140-122).
10. Collagen type I (Sigma-Aldrich, Cat. #C9791).
11. Dimethyl sulfoxide (DMSO, Sigma, Cat. #D-2650).

2.2 Laboratory Supplies

1. Tubes: 15 and 50 mL polypropylene (e.g., BD, Cat. #352097 and #352098).
2. Filters: disposable sterile 0.22 μm pore size and 500 mL sterile filter system (e.g., Corning, Cat. #430769).
3. Petri dishes: 35 and 65 mm (e.g., BD, Cat. #351008).
4. Cell strainer: disposable 70 μm pore size (e.g., BD, Cat. #352350).
5. Flasks: 25 cm^2 (T-25) and 75 cm^2 (T-75) (e.g., BD, Cat. #353109 and #353136).
6. Cryovials (1.5 mL, e.g., Nalgene, Cat. #5000-1020, see Note 3).
7. "Mr. Frosty" freezing container (Nalgene, Cat. #5100-0001) or empty 15 mL styrofoam boxes.

2.3 Required Equipment

1. Sterile surgical equipment (e.g., forceps, scissors).
2. Laminar flow tissue culture hood (e.g., The Baker Company, Model #SG 403).
3. Incubator to maintain 37°C, >95% humidity and an atmosphere of 5% CO_2 (e.g., HERAcell 150, Thermo Fisher, Cat. #51022393).
4. Refrigerated benchtop centrifuge (e.g., Legend RT, Thermo Fisher, Cat. #75004377).
5. Laboratory balance (e.g., Mettler Toledo, Model #PM4000).
6. Pipet-Aid (e.g., Drummond Scientific, Cat. #4-000-101).
7. Adjustable micropipettes: 10, 200 and 1,000 μL (e.g., Gilson, Model #P200N) with sterile disposable plastic pipette tips (e.g., Denville Scientific, Cat. #P1326-CPS).
8. Hemocytometer (e.g., Hausser 1475) or automatic cell counter (e.g., Invitrogen, Cat. #C10281).
9. Inverted light microscope with phase contrast capabilities and objectives of 5×, 10×, 20× magnification (e.g., Nikon, Model #TMS 215798).

3 Methods

3.1 Solution Preparation

1. Tissue digestion solution: The working concentrations are 0.2% (wt/vol) collagenase Type XI, 2.4 U/mL dispase, and 0.1% (wt/vol) Trypsin-EDTA. Both solutions are prepared in HBSS and filtered using a 0.22-μm sterile filter, aliquoted into 10 mL volume aliquots, and stored at −20°C. Solutions should be pre-warmed to 37°C before use.
2. Proliferation Medium (PM): Except for DMEM, all components of the PM should be preprepared and stored at −20°C. Pre-warm all the components to 37°C before use. To prepare 500 mL of PM, combine 392.5 mL DMEM, 50 mL FBS (10%), 50 mL HS (10%), and 5 mL P/S (1%) in a 500 mL sterile filter attached to the filter receiver. Add 2.5 mL CEE (0.5%) as the last 10 mL of solution is drawn through the sterile filter system, as it clogs the filter and retards the process. PM should be prepared fresh for each new culture of MDSPCs. Unused PM can be stored at 4°C for a maximum of 4 weeks.
3. Freezing medium: 1:10 dilution of DMSO: FBS or PM (see Note 4).
4. Collagen coating solution:

   Day 1:

   Under a laminar flow hood using sterile technique, add 0.1 g of collagen type I to a sterile 1 L glass bottle with a screw cap. In a sterile graduated cylinder, add 5.7 mL glacial acetic acid to 1 L of double-distilled water. Add the mixture to collagen by filtering the acid/water through a 500 mL bottle top sterilization filter. Stir with a magnetic stir bar overnight.

   Day 2:

   Under a laminar flow hood using sterile technique, pour collagen type I into 50 mL sterile polystyrene centrifuge bottles and spin at 1,932 × *g* for 30 min at 4°C. Transfer the supernatant to sterile (autoclaved) bottles under a laminar flow hood. Slowly add 50 mL of chloroform (caution: volatile) per 500 mL solution using a sterile glass funnel. The chloroform will form a layer under the collagen solution. DO NOT SHAKE OR STIR. Allow the solution to sit overnight (can be extended to 2 days) at 2–8°C.

   Day 3:

   Under a laminar flow hood using sterile technique, transfer the top aqueous layer (containing the collagen) to a new sterile 1 L bottle and store at 2–8°C. Discard the bottom layer containing the chloroform according to your institution's hazardous waste guidelines.

3.2 Collagen Coating Flasks and Plates

To coat the flasks or plates with collagen type I, warm the solution in 37°C water bath. Add the recommended volume of collagen to the flasks/plates (see Note 5). Make sure the solution is evenly distributed by gently rocking them until the surface is covered. Repeat frequently over the next 2 h. Keep the flasks/plates open in a culture hood with blower on for a minimum of 3–4 h, but ideally for 7 h. After this, pipette off the excess collagen solution to a flask labeled as "used collagen type I". This can be reused for future coating. Leave the flasks/plates partially open in a running laminar flow hood overnight to promote drying. Keep the flasks/plates in a laminar flow hood with the germicidal UV lamp on for 2 h. The next day, cap the flasks/plates, label, bag in the original wrapping and store at room temperature.

3.3 Isolating and Digesting Skeletal Muscle from Mice

1. Sterilize surgical tools.
2. Euthanize the animal by CO_2 asphyxiation.
3. Dissect out the skeletal muscle. For adult mice, use the hind limb muscles including gastrocnemius, soleus and quadriceps muscle (~350–450 mg, also see Note 6). Place the tissue from a single animal into a 15 mL tube containing sufficient HBSS to cover the tissue and store on ice.
4. Transfer one skeletal muscle at a time to a sterile 65 mm petri dish and wash it three times with sterile HBSS to remove any debris that may have adhered to the tissue during the dissection.
5. Using forceps and scissors, dissect the muscle from other tissues including the bone, tendon, nerve, major blood vessels, fat and connective tissue in a sterile 35 mm petri dish.
6. Add 10 mL of sterile HBSS to a fresh petri dish and begin the mechanical digestion of the trimmed muscle by mincing it into a coarse slurry using very sharp spring scissors.
7. Transfer the minced tissue slurry into a 15 mL conical tube using a 10 mL transfer pipette and centrifuge at $930 \times g$ at 2–8°C for 5 min.
8. Aspirate off the supernatant, resuspend the pellet with HBSS and centrifuge again as described in step 7.
9. Aspirate off the supernatant and weigh the pellet (slurry) using a precise scale. Zero the scale using an empty 15 mL conical tube.
10. Begin the enzymatic digestion by adding 10 mL of pre-warmed 0.2% collagenase type XI digestion solution (approximately 1 mL per 0.1 g muscle pellet) in a 15 mL conical tube. Incubate at 37°C. Gently rock the tube by hand every 15 min. After 1 h, centrifuge at $2,630 \times g$ for 5 min (see Note 7).

11. Aspirate off the supernatant and resuspend the pellet/slurry in 10 mL of pre-warmed dispase (2.4 U/mL). Incubate at 37°C for 45 min. Gently rock the tube by hand every 15 min. Centrifuge at 2,630 × *g* for 5 min.
12. Aspirate off the supernatant and resuspend the viscous slurry in 0.1% Trypsin-EDTA solution diluted in HBSS. Incubate at 37°C for 30 min. Gently rock the tube by hand every 15 min. Centrifuge at 2,630 × *g* for 5 min.
13. Aspirate out the supernatant and resuspend the cell pellet in 10 mL of PM.
14. Pipette the resuspended pellet through a 70 μm cell strainer placed on top of a 50 mL sterile conical tube. Rinse the strainer with PM to collect all of the cells.

3.4 Pre-plating Steps (Fig. 1)

1. Plate the cells from one animal/tube on one collagen type I-coated T-25 flask and incubate at 37°C in a humidified, 5% CO_2 incubator for 2 h.
2. Two hours after plating, early adhering cells (mainly fibroblasts) will attach to the surface of the flask; label this cell population as pre-plate 1 (pp1).

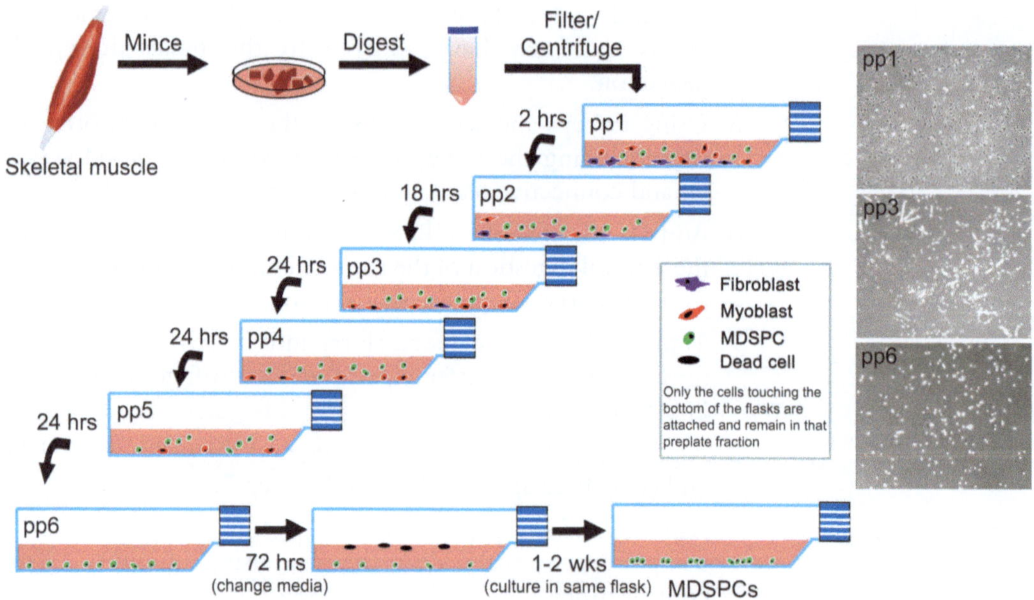

Fig. 1 Isolation of muscle-derived stem/progenitor cells (MDSPCs). Skeletal muscle is mechanically and enzymatically dissociated to a single cell suspension. The muscle cells are resuspended in proliferation medium and plated on collagen type I-coated flasks. Different populations of muscle-derived cells are isolated based on their adhesion characteristics. Pre-plate 1 (pp1) cells adhere in the first 2 h after isolation and consist predominantly of primary fibroblasts. Subsequent pre-plates obtained at 18–24 h intervals (pp2–5), contain a mixture of fibroblast and myoblast-like cells. Cells in pp6 take an additional 72 h to attach. Most of the cells from pp6 die during the first 1–2 weeks of culturing. The few adherent cells that survive proliferate to form colonies. These viable, proliferating pp6 cells are called MDSPCs

3. Transfer the media containing the nonadherent cells into a second collagen type I-coated T-25 flask labeled pp2. Return the flasks to the incubator.
4. After 18 h, transfer the media from pp2 into a sterile 15 mL conical tube and centrifuge at 930 × g at for 5 min. Aspirate out the supernatant and resuspend the cell pellet in 5 mL of PM and transfer the media containing the nonadherent cells into a third T-25 flask labeled pp3. Add 5 mL of fresh PM to the pp2 flask. Return the flasks to the incubator.
5. Repeat step 4 three times to obtain pp6. This last cell suspension can be maintained for up to 72 h to maximize adherence of the slowest adhering cells. Change the media on pp2–5 cells every other day and freeze as soon as a sufficient number is reached. Most of the cells in pp6 die, but the surviving cells will slowly begin to proliferate, creating small adherent colonies. These viable cells are small, round and refractive. The surviving cells at pp6 are termed muscle derived stem/progenitor cells (MDSPCs, see Note 8).

3.5 Expanding MDSPCs (Fig. 2)

1. Remove the culture medium (PM) and wash the cells with HBSS.
2. Apply pre-warmed 0.25% Trypsin-EDTA in HBSS and gently tap the corners or bottom of the flasks to stimulate cell detachment. After ~1 min, check the flask under the microscope to ensure that all of the cells are detached.

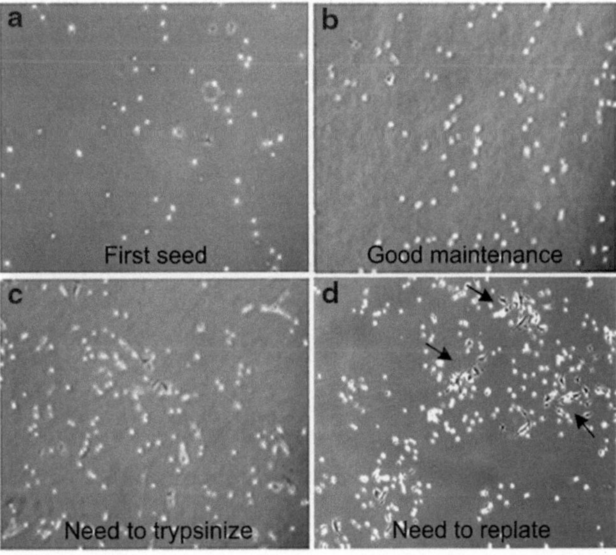

Fig. 2 MDSPC morphology and maintenance. Representative images indicating (**a**) the initial density of MDSPCs yielded at pp6 (~200 cells/cm^2). (**b**) MDSPCs maintained at an optimal confluence of <30%. (**c**) MDSPCs in need of trypsinization and splitting 1:2. (**d**) MDSPCs ready to be replated to eliminate the fibroblast-like cells (*arrows*). Images are at 5× magnification

3. Inhibit the Trypsin-EDTA by adding PM (~3 mL for T-25 flasks). Transfer the cell suspension to a 15 mL conical tube and centrifuge at 930 × *g* for 5 min at 4°C.
4. Aspirate off the supernatant and resuspend the pellet in PM. Count the cells and plate at a density of 500–1,000 cells/cm^2 (e.g., ~12,000–25,000 cells in T-25 flask, see Note 9) in a fresh collagen type I-coated flask. Incubate the cells at 37°C in a humidified, 5% CO_2 incubator. Maintain the cells at a low cell density (30% confluence) to minimize cell differentiation. MDSPCs need to be trypsinized when small colonies are visible regardless of the overall cell density. MDSPCs usually remain quiescent after isolation; therefore, 3–4 weeks are needed to obtain sufficient cells for cryopreservation or experimentation (see Note 10).

3.6 Replating MDSPCs (Fig. 2)

If pp6 cells contain a high percentage of large, flat, fibroblast-like cells, it is necessary to replate the cells as described below.

1. Follow steps 1–3 in Subheading 3.5.
2. Aspirate off the supernatant and resuspend the pellet in PM, then plate the cells in a fresh collagen type I-coated T-25 flask. Allow the cells to adhere to the flask for 20–30 min in the incubator.
3. Collect the media containing the non-adherent cells and "replate" them onto a new collagen type I-coated T-25 flask. Incubate the cells at 37°C in a humidified, 5% CO_2 incubator. Replating steps can be repeated several times if necessary.

3.7 Freezing MDSPCs

Freeze each preplate population (pp1–pp5) when the cells are 80% confluent. The pp6 cells can be frozen once a sufficient number has been reached, while keeping the cells at low confluence (<30%) at all times. Cell pellets containing $\sim 1 \times 10^5 - 3 \times 10^5$ cells should be mixed with freezing media (1:10 dilution of DMSO: FBS) and placed in −20°C for 2 h. The vials are then placed in the middle of two 15 mL styrofoam boxes or a Mr. Frosty freezing container then transferred to a −80°C freezer. For long term preservation, the cells should be transferred to liquid nitrogen (see Note 11).

3.8 Thawing MDSPCs

1. Transfer frozen vials of cells from the −80°C freezer or liquid nitrogen to the laboratory on dry ice.
2. Thaw the content of the vials using a 37°C water bath. Thawing needs to be done quickly (<1 min) because DMSO is toxic to the cells.
3. Transfer the content of the vial into a 15 mL conical tube containing 5 mL of PM and centrifuge at 930 × *g* for 5 min at 4°C. Aspirate off the supernatant. Resuspend the cell pellet in a small volume of PM media. The contents of one cryovial

($\sim 1 \times 10^5$/T-75) should be plated according to step 4 in Subheading 3.5.

4. The day after thawing, change the media to remove the dead cells.

3.9 Measuring the Quality/Function of Isolated MDSPCs (Table 1)

After isolation, the quality of MDSPCs is assessed by measuring stem-like properties including expression of stem/progenitor markers self-renewal and multi-lineage differentiation (summarized in Table 1).

4 Notes

1. Using HBSS with Ca^{2+} and Mg^{2+} reduces the enzymatic activity of trypsin, but it is still effective for gently releasing stem/progenitor cells from plates.
2. In our experience, CEE is critical for the isolation and maintenance of MDSPCs. CEE from different manufacturers and even different lots produced by the same manufacturer have a profound effect on the quality of MDSPCs including their capacity for proliferation and differentiation. The best results are obtained if the CEE is not precleared by centrifugation or ultrafiltration.
3. Ideally, MDSPCs should be stored submerged in liquid nitrogen. Note that this requires special cryovials that can withstand the pressure changes that occur when cells are thawed. It is also important not to overfill the cryovials to prevent the tubes from exploding during the thawing process.
4. Viability of MDSPCs after thawing is dramatically improved if a 1:10 (vol/vol) ratio of DMSO:FBS is used rather than DMSO:PM.
5. Amount of collagen type I needed for coating of flasks/plates:

T25	2 mL
T75	3.5–4 mL
T175	7.5–8 mL
6-well plate	1 mL/well
12-well plate	0.5 mL/well
24-well plate	0.25 mL/well
96-well plate	50 μL/well

6. For newborn animals, both the hindlimb and forelimb muscles are collected.
7. Collagenase type XI and dispase digestion can be extended when a larger amount of muscle must be digested. But the cells are sensitive to prolonged exposure to Trypsin-EDTA.

Table 1
Endpoints used to evaluate the quality and function of MDSPCs and how those parameters change in cells isolated from aged organisms

	Assay used	Unit	References	Young murine WT-MDSPCs	Aged murine MDSPCs
Marker profile Sca1+ CD34+ CD45−	Immuno; cytochemistry; RT-PCR	% Cells of total population	(21)	18% of cells at pp3 (host 3 weeks old) 9% at pp3 (host 4–5 months old)	3% of cells at pp3 (host 3 years old)
Proliferation	Live-cell imaging	DT (h) PDT (h)	(21)	DT = 13 ± 2 h; PDT = 15 h (host 3 weeks old)	DT = 16 ± 6 h; PDT = 18 h (host 2 years old)
Differentiation	Myogenic, osteogenic, chondrogenic, adipogenic	± Differentiation capacity	(13, 21)	Positive for all four lineages	Adipocyte > myocyte > chondrocyte > osteocyte
Engraftment potential and regeneration	IM injection into *mdx*/SCID mice or WT mice after cardiotoxin injury	Number or % of regenerating myofibers and their cross-sectional area	(13, 21, 61)	70% of regenerating, centronucleated fibers have an area >1,000 μm^2	70% of regenerating, centronucleated fibers have an area <750 μm^2

MDSPCs are double positive for Sca1 and CD34 and negative for CD45 (13, 21, 34). Although, variability and alteration in the expression levels has been observed as a consequence of in vitro culturing (55). Growth kinetics (proliferation and/or self-renewal) is measured using a live-cell imaging (LCI) system. This system allows for time-lapsed imaging of single cells or colonies over long periods of time, so that growth and differentiation kinetics can be accurately measured (20, 56, 57). To evaluate the differentiation potential of MDSPCs, the cells are switched to specific differentiation media for myogenic, osteogenic, chondrogenic, and adipogenic lineages. Subsequently, the cultures are analyzed for lineage-specific morphology and differentiation markers (21). The function (regenerative capacity) of MDSPCs is evaluated in vivo by injecting the cells into the gastrocnemius muscle of *mdx*/SCID mice, a mouse model of Duchenne muscular dystrophy with profound muscle degeneration due to lack of dystrophin expression (58) that are also immunocompromised (SCID). Two weeks after injecting MDSPCs, the muscle is harvested and stained for dystrophin to detect donor-derived myofibers. Alternatively, muscle regeneration can be tested in wild-type mice following cardiotoxin injury (21, 59). In this case, regenerating fibers are identified by centronucleation (21, 60)

DT division time, *IM* intramuscular injection, *PDT* population-doubling time, *WT* wild-type

8. If contamination is observed, discard all cultures and clean the hood and incubators thoroughly with a topical application of disinfectant (Coflikt, Decon Laboratories Cat. #4101). Autoclave all removable parts of the hood and incubator before starting a new isolation. It is not recommended to increase the dose of antibiotics or to use antimycotics since this may affect the stemness properties of MDSPCs.
9. The seeding density is critical and varies depending on the cell proliferation kinetics. Low cell density impedes cell proliferation and high density causes the cells to fuse (differentiate). The optimal density must be determined empirically (Fig. 2). Cell proliferation rates vary depending on the sex, age and health of the donor as well as the amount of muscle tissue available for cell isolation. Consistency in cell isolation and culturing of MDSPCs is the key to maintaining pluripotency.
10. Since MDSPCs are quiescent cells, it takes a minimum of 1–2 weeks for cells to show any sign of proliferation. This time varies depending upon the number of cells isolated at pp6. Trypsinization should be avoided until visible colonies are evident.
11. MDSPCs may be stored at −80°C in DMSO:FBS for short term (up to 3 months). Long term storage (>3 months) requires submersion in liquid nitrogen to maximize cell survival and yield.

Acknowledgements

This work was supported by multiple institutes throughout many years including the National Institutes of Health, the US Department of Defense, the Muscular Dystrophy Association, the University of Pittsburgh Cancer Institute, and generous support from the William F. and Jean W. Donaldson Endowed Chair at the Children's Hospital of Pittsburgh of UPMC, and the Henry J. Mankin Endowed Chair for Orthopaedic Research at the University of Pittsburgh.

References

1. Pera MF (2001) Scientific considerations relating to the ethics of the use of human embryonic stem cells in research and medicine. Reprod Fertil Dev 13(1):23–29
2. Hook CC (2010) In vitro fertilization and stem cell harvesting from human embryos: the law and practice in the United States. Pol Arch Med Wewn 120(7–8):282–289
3. Blendon RJ et al (2011) The public, political parties, and stem-cell research. N Engl J Med 365(20):1853–1856
4. Usas A et al (2011) Skeletal muscle-derived stem cells: implications for cell-mediated therapies. Medicina (Kaunas) 47(9):469–479
5. Mizuno H (2010) Adipose-derived stem and stromal cells for cell-based therapy: current status of preclinical studies and clinical trials. Curr Opin Mol Ther 12(4):442–449
6. Wilson A et al (2011) Adipose-derived stem cells for clinical applications: a review. Cell Prolif 44(1):86–98
7. Carr LK et al (2008) 1-year follow-up of autologous muscle-derived stem cell injection pilot study to treat stress urinary incontinence. Int Urogynecol J Pelvic Floor Dysfunct 19(6):881–883
8. Owen M (1988) Marrow stromal stem cells. J Cell Sci Suppl 10:63–76

9. da Silva Meirelles L et al (2006) Mesenchymal stem cells reside in virtually all post-natal organs and tissues. J Cell Sci 119(Pt 11): 2204–2213
10. Watts C et al (2005) Anatomical perspectives on adult neural stem cells. J Anat 207(3):197–208
11. Lim DA et al (2007) The adult neural stem cell niche: lessons for future neural cell replacement strategies. Neurosurg Clin N Am 18(1): 81–92, ix
12. Baroffio A et al (1996) Identification of self-renewing myoblasts in the progeny of single human muscle satellite cells. Differentiation 60(1):47–57
13. Qu-Petersen Z et al (2002) Identification of a novel population of muscle stem cells in mice: potential for muscle regeneration. J Cell Biol 157(5):851–864
14. Sherwood RI et al (2004) Isolation of adult mouse myogenic progenitors: functional heterogeneity of cells within and engrafting skeletal muscle. Cell 119(4):543–554
15. Shen W et al (2003) Adipose tissue quantification by imaging methods: a proposed classification. Obes Res 11(1):5–16
16. Toma JG et al (2001) Isolation of multipotent adult stem cells from the dermis of mammalian skin. Nat Cell Biol 3(9):778–784
17. Burke ZD et al (2007) Stem cells in the adult pancreas and liver. Biochem J 404(2):169–178
18. Barker N et al (2007) Identification of stem cells in small intestine and colon by marker gene Lgr5. Nature 449(7165):1003–1007
19. Yen TH, Wright NA (2006) The gastrointestinal tract stem cell niche. Stem Cell Rev 2(3): 203–212
20. Deasy BM et al (2005) Long-term self-renewal of postnatal muscle-derived stem cells. Mol Biol Cell 16(7):3323–3333
21. Lavasani M et al (2012) Muscle-derived stem/progenitor cell dysfunction limits healthspan and lifespan in a murine progeria model. Nat Commun 3:608
22. Yokoyama T et al (2001) Autologous primary muscle-derived cells transfer into the lower urinary tract. Tissue Eng 7(4):395–404
23. Chancellor MB et al (2001) Gene therapy strategies for urological dysfunction. Trends Mol Med 7(7):301–306
24. Musgrave DS et al (2002) Human skeletal muscle cells in ex vivo gene therapy to deliver bone morphogenetic protein-2. J Bone Joint Surg Br 84(1):120–127
25. Sakai T et al (2002) The use of ex vivo gene transfer based on muscle-derived stem cells for cardiovascular medicine. Trends Cardiovasc Med 12(3):115–120
26. Peng H et al (2002) Synergistic enhancement of bone formation and healing by stem cell-expressed VEGF and bone morphogenetic protein-4. J Clin Invest 110(6):751–759
27. Ikezawa M et al (2003) Dystrophin delivery in dystrophin-deficient DMDmdx skeletal muscle by isogenic muscle-derived stem cell transplantation. Hum Gene Ther 14(16):1535–1546
28. Hannallah D et al (2004) Retroviral delivery of Noggin inhibits the formation of heterotopic ossification induced by BMP-4, demineralized bone matrix, and trauma in an animal model. J Bone Joint Surg Am 86-A(1):80–91
29. Deasy BM et al (2009) Effect of VEGF on the regenerative capacity of muscle stem cells in dystrophic skeletal muscle. Mol Ther 17(10): 1788–1798
30. Peng H et al (2005) VEGF improves, whereas sFlt1 inhibits, BMP2-induced bone formation and bone healing through modulation of angiogenesis. J Bone Miner Res 20(11): 2017–2027
31. Chirieleison SM et al (2012) Human muscle-derived cell populations isolated by differential adhesion rates: phenotype and contribution to skeletal muscle regeneration in Mdx/SCID mice. Tissue Eng Part A 18(3–4):232–241
32. Okada M et al (2012) Human skeletal muscle cells with a slow adhesion rate after isolation and an enhanced stress resistance improve function of ischemic hearts. Mol Ther 20(1):138–145
33. Torrente Y et al (2001) Intraarterial injection of muscle-derived CD34(+)Sca-1(+) stem cells restores dystrophin in mdx mice. J Cell Biol 152(2):335–348
34. Lee JY et al (2000) Clonal isolation of muscle-derived cells capable of enhancing muscle regeneration and bone healing. J Cell Biol 150(5):1085–1100
35. Wright V et al (2002) BMP4-expressing muscle-derived stem cells differentiate into osteogenic lineage and improve bone healing in immunocompetent mice. Mol Ther 6(2):169–178
36. Kuroda R et al (2006) Cartilage repair using bone morphogenetic protein 4 and muscle-derived stem cells. Arthritis Rheum 54(2):433–442
37. Aguiari P et al (2008) High glucose induces adipogenic differentiation of muscle-derived stem cells. Proc Natl Acad Sci USA 105(4):1226–1231
38. Arriero M et al (2004) Adult skeletal muscle stem cells differentiate into endothelial lineage and ameliorate renal dysfunction after acute ischemia. Am J Physiol Renal Physiol 287(4): F621–F627
39. Jackson KA et al (1999) Hematopoietic potential of stem cells isolated from murine skeletal muscle. Proc Natl Acad Sci USA 96(25):14482–14486
40. Romero-Ramos M et al (2002) Neuronal differentiation of stem cells isolated from adult muscle. J Neurosci Res 69(6):894–907

41. Vourc'h P et al (2004) Isolation and characterization of cells with neurogenic potential from adult skeletal muscle. Biochem Biophys Res Commun 317(3):893–901
42. Winitsky SO et al (2005) Adult murine skeletal muscle contains cells that can differentiate into beating cardiomyocytes in vitro. PLoS Biol 3(4):e87
43. Arsic N et al (2008) Muscle-derived stem cells isolated as non-adherent population give rise to cardiac, skeletal muscle and neural lineages. Exp Cell Res 314(6):1266–1280
44. Tamaki T et al (2002) Identification of myogenic-endothelial progenitor cells in the interstitial spaces of skeletal muscle. J Cell Biol 157(4):571–577
45. Cao B et al (2003) Muscle stem cells differentiate into haematopoietic lineages but retain myogenic potential. Nat Cell Biol 5(7):640–646
46. Lee JY et al (2001) Effect of bone morphogenetic protein-2-expressing muscle-derived cells on healing of critical-sized bone defects in mice. J Bone Joint Surg Am 83-A(7):1032–1039
47. Matsumoto T et al (2009) Cartilage repair in a rat model of osteoarthritis through intraarticular transplantation of muscle-derived stem cells expressing bone morphogenetic protein 4 and soluble Flt-1. Arthritis Rheum 60(5):1390–1405
48. Oshima H et al (2005) Differential myocardial infarct repair with muscle stem cells compared to myoblasts. Mol Ther 12(6):1130–1141
49. Payne TR et al (2005) Regeneration of dystrophin-expressing myocytes in the mdx heart by skeletal muscle stem cells. Gene Ther 12(16):1264–1274
50. Lee JY et al (2003) The effects of periurethral muscle-derived stem cell injection on leak point pressure in a rat model of stress urinary incontinence. Int Urogynecol J Pelvic Floor Dysfunct 14(1):31–37, discussion 37
51. Cannon TW et al (2003) Improved sphincter contractility after allogenic muscle-derived progenitor cell injection into the denervated rat urethra. Urology 62(5):958–963
52. Chermansky CJ et al (2004) Intraurethral muscle-derived cell injections increase leak point pressure in a rat model of intrinsic sphincter deficiency. Urology 63(4):780–785
53. Kwon D et al (2006) Periurethral cellular injection: comparison of muscle-derived progenitor cells and fibroblasts with regard to efficacy and tissue contractility in an animal model of stress urinary incontinence. Urology 68(2):449–454
54. Gussoni E et al (1999) Dystrophin expression in the mdx mouse restored by stem cell transplantation. Nature 401(6751):390–394
55. Jankowski RJ et al (2001) Flow cytometric characterization of myogenic cell populations obtained via the preplate technique: potential for rapid isolation of muscle-derived stem cells. Hum Gene Ther 12(6):619–628
56. Sherley JL et al (1995) A quantitative method for the analysis of mammalian cell proliferation in culture in terms of dividing and non-dividing cells. Cell Prolif 28(3):137–144
57. Deasy BM et al (2003) Modeling stem cell population growth: incorporating terms for proliferative heterogeneity. Stem Cells 21(5):536–545
58. Hoffman EP et al (1987) Conservation of the Duchenne muscular dystrophy gene in mice and humans. Science 238(4825):347–350
59. d'Albis A et al (1988) Regeneration after cardiotoxin injury of innervated and denervated slow and fast muscles of mammals. Myosin isoform analysis. Eur J Biochem 174(1):103–110
60. Matsuura T et al (2007) Skeletal muscle fiber type conversion during the repair of mouse soleus: potential implications for muscle healing after injury. J Orthop Res 25(11): 1534–1540
61. Lavasani M et al (2006) Nerve growth factor improves the muscle regeneration capacity of muscle stem cells in dystrophic muscle. Hum Gene Ther 17(2):180–192

Chapter 6

Human Myoblasts from Skeletal Muscle Biopsies: In Vitro Culture Preparations for Morphological and Cytochemical Analyses at Light and Electron Microscopy

Manuela Malatesta, Marzia Giagnacovo, Rosanna Cardani, Giovanni Meola, and Carlo Pellicciari

Abstract

We describe protocols for the isolation of satellite cells from human muscle biopsies, for the in vitro culture of proliferating and differentiating myoblasts, and for the preparation of cell samples suitable for morphological and cytochemical analyses at light and electron microscopy. The procedures described are especially appropriate for processing small muscle biopsies, and allow obtaining myoblast/myotube monolayers on glass coverslips, thus preserving good cell morphology and immunoreactivity for protein markers of myoblast proliferation, differentiation, and senescence.

These cell preparations are suitable for cytochemical, immunocytochemical, and FISH procedures at light microscopy, and can be observed not only in bright field, phase contrast, and differential interference contrast but also in fluorescence (which can hardly be used for cells grown on conventional plastic surfaces, which generally exhibit intense autofluorescence). In their ultrastructural cytochemical application, the protocols are intended for post-embedding techniques, by which ultrathin sections from a single sample may be used for detecting a wide variety of molecular markers.

Keywords Myoblasts, In vitro proliferation, In vitro differentiation, In vitro senescence, Light microscopy, Transmission electron microscopy, Morphology, Cytochemistry

1 Introduction

In recent years, cytochemistry and immunocytochemistry have become a widely used approach to investigate the structural organization and function of skeletal muscle cells, especially as diagnostic tools in neuromuscular disorders or in the aged-related muscle loss, sarcopenia (e.g., refs. 1–4). In particular, in vitro cultured myoblasts derived from satellite cells provide a useful and reliable model for studying muscular precursor cells, and can be used to elucidate the basic mechanisms involved in normal differentiation or in muscle pathogenesis. However, the bioptic pathological

Kursad Turksen (ed.), *Stem Cells and Aging: Methods and Protocols*, Methods in Molecular Biology, vol. 976, DOI 10.1007/978-1-62703-317-6_6,

material suitable for in vitro studies can often be collected in small amounts especially when patient's skeletal muscle mass is reduced, with consequently decreased quantity of satellite cells exhibiting depletion in their activation, proliferation, and differentiation capabilities (as it usually occurs with dystrophic or sarcopenic patients). Under these circumstances, the application of in vitro culture techniques allows to expand the original cell population, thus optimizing the use of limited cellular stocks. It is, however, well known that cell primary cultures from non-tumor tissues have limited survival in vitro, so that necessarily it is almost impossible to obtain large amount of cells to be studied.

The histochemical approach represents a unique and irreplaceable tool for investigating the proliferation potential and differentiation capabilities of myoblasts derived from satellite cells: actually, the examination by light or electron microscopy of histochemically labeled samples allows to relate the expression of specific molecular markers to the cytomorphological features of proliferating or differentiating myoblasts.

Differentiating myoblasts give rise to the formation of the so-called myotubes, in a process that entails both the myoblast fusion and the intracellular restructuring of the cytoskeletal proteins to give rise to myofibril formation. To investigate in detail this progressive reorganization, it is essential to preserve the original cell morphology: cell detachment from the surface of the culture flasks or dishes should therefore be avoided, and cell monolayers must be preserved in their integrity for microscopic examination.

In the present work we present protocols for the isolation of human satellite cells, the in vitro culture of myoblasts and the preparation of cell samples for morphological and cytochemical analyses at light and electron microscopy; the described procedure is especially suitable for processing small muscle biopsies and, in our experience, the obtained myogenic index (i.e., the percentage of myoblasts in the cell population) ranges between 50 and 80, in the cell cultures at the first passage. Among the manifold cytochemical and immunocytochemical techniques which can be used to investigate satellite-cell-derived myoblasts in vitro, we decided to focus our attention on protocols for detecting protein markers of myoblast proliferation, differentiation, and senescence. Moreover, for ultrastructural cytochemistry (immunocytochemistry and in situ hybridization included), the protocols are intended for post-embedding techniques, which allow to use ultrathin sections from a single sample for the detection of several different molecular markers (5).

These protocols have been widely used in our laboratories, and some papers have recently been published on myoblasts obtained from patients affected by myotonic dystrophy which is characterized by severe muscle fiber disorganization and loss (6–9).

2 Materials

Prepare all solutions using sterile ultrapure water (18 MΩ cm at 25°C) and analytical grade reagents. Follow scrupulously all the instructions for toxic material use, and the rules for waste disposal.

2.1 Reagents for Myoblast Isolation and In Vitro Culture

1. *Dissection buffer, pH 7.5*: prepare 100 mL stock solutions using sterile water and sterile glass beakers; for NaCl stock solution, dissolve 760 mg NaCl; for KCl stock solution, dissolve 22.4 mg KCl; for $NaHPO_4$ stock solution, dissolve 13.8 mg $NaHPO_4 \times H_2O$; for Hepes stock solution, dissolve 715 mg Hepes. Working solution: pour into a sterile beaker 10 mL of the NaCl stock solutions, 0.6 mL of the KCl, 2 mL of the $NaHPO_4$, 30 mL of the Hepes one, and add 180 mg glucose, while stirring; when dissolved, add water to a volume of 100 mL and, if necessary, adjust the pH to 7.5. Aliquots of the working solution may be stored for several months at -20°C.
2. *Trypsin-EDTA solution*: 0.05% Trypsin, 0.02% EDTA in PBS without Calcium, Magnesium and Phenol Red (Euroclone, Milan, Italy) (see Note 2).
3. *Fetal Bovine Serum* (FBS; Euroclone).
4. *Dexamethasone stock solution*: prepare, under sterile conditions, a solution of 25 mg dexamethasone in 1 mL of methanol. Dilute 1 mL of this stock solution in 62.7 mL of sterile water (0.8 mM dexamethasone, working solution) (see Note 2).
5. *Human Epidermal Growth Factor (EGF, recombinant, expressed in E. coli) stock solution*: prepare, under sterile conditions, a solution of 60 μL of acetic acid in 100 mL of sterile water; add 0.1% bovine serum albumin (BSA), then add 1 mg EGF (EGF working solution) (see Note 2).
6. *Proliferative medium, pH 7.4*: F-10 medium solution in 1 L of sterile water, dissolve 9.8 g Nutrient Mixture F-10 with L-Glutamine and without $NaHCO_3$ (Sigma-Aldrich), and 1.2 g $NaHCO_3$ (see Note 1); in 850 mL of this F10 medium solution, dissolve 500 mg of BSA, 500 mg of cell-culture-tested fetuin from fetal calf serum, 3 g of glucose, 990 μL of dexamethasone stock solution, 1 mL of EGF stock solution, and finally 500 mg of insulin from bovine pancreas previously dissolved in a few drops of 0.1N HCl (all these reagents are from Sigma-Aldrich). If necessary, adjust the pH to 7.4 using drops of 0.5N NaOH. Sterilize the final solution through 0.22 μm mesh cellulose filters (VWR international, Milan, Italy); aliquots may be stored at -20°C for several months. Immediately before use, thaw and add to 84 mL of proliferative medium 15 mL of FBS (Euroclone), and 1 mL of penicillin/streptomycin solution (Euroclone) containing 10,000 U/mL of penicillin and 10 mg/mL of streptomycin.

7. *Phosphate buffer saline* (PBS), *pH 7.4*: prepare a 10× solution by dissolving 2.45 g KH_2PO_4, 14.63 g $Na_2HPO_4 \times 2H_2O$, and 80 g NaCl in 900 mL of water (see Note 1); when the solutes are completely dissolved, add water to 1 L volume, and store at 4°C. Before use, dilute this 10× solution 1:10 in water to obtain 0.1 M PBS working solution (for sake of simplicity, in the following text this will be referred to as PBS only).
8. *Dimethyl sulfoxide* (DMSO; Sigma-Aldrich).
9. *Isopropyl alcohol* (Sigma-Aldrich).

2.2 Reagents for Myoblast and Myotube Preparation on Glass Coverslips

1. *PBS, pH 7.4*: prepare as described in Subheading 2.1, item 7.
2. *Trypsin-EDTA solution*: *see* Subheading 2.1, item 2.
3. *Proliferative medium*: prepare as described in Subheading 2.1, item 6.
4. *Trypan blue solution*: prepare a 1% stock solution in PBS; store in a dark bottle (to be filtered after prolonged storage). Before use dilute 1:100 with PBS.
5. *Differentiation medium, pH 7.4*: add 0.01 mg/mL insulin (prepared as described above) to high-glucose Dulbecco's Modified Eagle's Medium (DMEM) (Euroclone). Adjust the pH to 7.4 with drops of 0.5N NaOH, and sterilize (see Subheading 2.1, item 6, and Note 2). Before use, add to 92 mL of the DMEM-insulin solution 7 mL of FBS and 1 mL of penicillin/streptomycin solution (see Subheading 2.1, item 6).

2.3 Reagents for Detection of Proliferation and Senescence Markers at Light Microscopy

1. *5-Bromo-2′-deoxyuridine (BrdU) containing proliferation medium*: to prepare a stock solution, dissolve 12.2 mg BrdU (Sigma-Aldrich) in 10 mL of Nutrient Mixture F-10 with L-Glutamine and without $NaHCO_3$ (see Note 2). Before use, thaw and dilute the stock solution 1:1,000 in pre-warmed proliferative medium.
2. *70% ethanol*: dilute the ethanol in water and cool at −20°C.
3. *PBS, pH 7.4*: prepare as described in Subheading 2.1, item 7.
4. *2N HCl*: add 16.8 mL of 37% HCl to water, to a volume of 100 mL.
5. *0.1 M sodium tetraborate (pH 8.2)*: dissolve 3.814 g sodium tetraborate in 100 mL of water (see Note 1).
6. *Anti-desmin antibody*: anti-desmin mouse monoclonal antibody (Clone D33, Dako, Glostrup, Denmark); immediately before use dilute 1:100 in PBS containing 0.1% BSA (Sigma-Aldrich) and 0.05% Tween-20 (Sigma-Aldrich).
7. *Alexa 594-labeled goat antibody recognizing mouse IgG* (Molecular Probes, Invitrogen, Paisley, UK): immediately before use, dilute 1:200 in PBS.

8. *Anti-BrdU antibody*: FITC-conjugated anti-BrdU mouse monoclonal antibody (Clone B44, BD Biosciences, Franklin Lakes, NJ); immediately before use dilute 1:20 in PBS containing 0.1% BSA (Sigma-Aldrich) and 0.05% Tween-20 (Sigma-Aldrich).
9. *Hoechst 33258* (Sigma-Aldrich): prepare a stock solution (0.1 mg/mL in water); store at 4°C. Before use, dilute 1:1,000 in PBS.
10. *Mounting medium*: immediately before use, mix PBS and glycerol 1:1.
11. *25% glutaraldehyde solution* (Sigma-Aldrich).
12. *Fixative solution for Senescence-Associated β-galactosidase (SA-β-Gal) staining*: dissolve 4 g paraformaldehyde in 80 mL of PBS (see Note 3), add 0.8 mL of 25% glutaraldehyde solution (see Note 4) and finally add PBS to a volume of 100 mL. If necessary, clarify with drops of 0.5N NaOH and adjust the pH to 7.4 (see Note 5).
13. *SA-β-Gal staining solution* (10): prepare the X-Gal stock solution by diluting 20 mg 5-bromo-4-chloro-3-indolyl β-D-galactoside (Sigma-Aldrich) in 1 mL of dimethylformamide (see Note 4). Prepare 10 mL of the following stock solutions using sterile water and sterile glass beakers: citric acid/NaH_2PO_4 stock solution (dissolve 768 mg citric acid and 552 mg $NaH_2PO_4 \times H_2O$, and adjust the pH to 6.0), $K_4Fe(CN)_6$ stock solution (dissolve 211 mg $K_4Fe(CN)_6 \times 3H_2O$), $K_3Fe(CN)_6$ stock solution (dissolve 164 mg $K_3Fe(CN)_6$), NaCl stock solution (dissolve 877 mg NaCl), $MgCl_2$ stock solution (dissolve 41 mg $MgCl_2 \times 6H_2O$). In a 35 mm petri dish, add 100 μL of all stock solutions, 50 μL of X-Gal solution and 450 μL of PBS (see Note 6).

2.4 Reagents for Transmission Electron Microscopy

1. *PBS, pH 7.4*: prepare as described in Subheading 2.1, item 7.
2. *25% glutaraldehyde solution* (Sigma-Aldrich).
3. *Fixative solution for ultrastructural morphology*: dissolve 2 g paraformaldehyde in 80 mL of PBS (see Note 3), add 10 mL of 25% glutaraldehyde solution (see Note 4) and finally add PBS to a volume of 100 mL.
4. *Fixative solution for ultrastructural cytochemistry*: dissolve 4 g paraformaldehyde in 100 mL of PBS (see Note 3). If necessary, clarify with drops of 0.5N NaOH and adjust the pH to 7.4 (see Note 5).
5. *1% Osmium tetroxide*: dissolve 2 g OsO_4 in 100 mL water (stock solution) (see Note 7). Before use dilute 1:1 with PBS.
6. *60% Acetone*: dilute acetone in water, store at room temperature (RT).

7. *EMbed-812 resin*: put 30.8 mL of Dodecenyl Succinic Anhydride (DDSA), 45.6 mL of EMbed-812, and 23.6 mL of Methyl 5-Norbornene-2,3-Dicarboxylic Anhydride (NMA) (Electron Microscopy Sciences, Hatfield, PA) in a 100 mL graduate glass cylinder (see Note 8); transfer to a glass or plastic beaker, and gently mix with a magnetic stirrer for about 10 min, then add 2 mL of 2,4,6-Tris(dimethylaminomethyl) phenol (DMP-30) (Electron Microscopy Sciences) and gently mix for at least 30 min (see Note 9).
8. *0.5 M NH4Cl solution*: dissolve 2.67 g NH_4Cl in 100 mL of PBS.
9. *Ethanol solutions*: 30, 50, 70 and 90% ethanol in water.
10. *LR White resin*: the resin is ready for use as it is provided by the supplier (Electron Microscopy Sciences).

3 Methods

3.1 Myoblast Isolation from Skeletal Muscle Biopsies and In Vitro Culture

All the procedures must be carried out under strictly sterile conditions, always using sterile plastics and pre-warmed sterile solutions at 37°C.

1. Use 3–4 mm^3 bioptic skeletal muscle samples.
2. Place the biopsy in a 60 mm petri dish containing 4 mL of dissection buffer.
3. Trim off blood vessels, fat and connective tissue by using small tweezers and scissors, mince into smaller pieces, and transfer them to a 35 mm petri dish containing 2 mL of dissection buffer.
4. Dissociate the muscle fibers using small tweezers and cut them repeatedly by scissors.
5. Gently aspirate the suspension of dissociated tissue with a 5 mL sterile pipette, and pour into a 15 mL tube; let the fragments sediment for 5 min, and then gently remove the liquid (which must be discarded) with a sterile pipette.
6. Add 6 mL of disaggregating trypsin solution and let the enzyme work for 15 min in a water bath at 37°C (see Note 10).
7. Let the suspension sediment, then aspirate about 5 mL of the liquid containing the isolated cells (see Note 11); pour the cell suspension into a 50 mL tube, and stop trypsin digestion by adding 0.5 mL of FBS.
8. Add 5 mL of disaggregating trypsin solution to the remaining tissue fragments and repeat passages 6 and 7 for further three times, in order to obtain the almost complete dissociation of the tissue fragments.

9. Centrifuge the collected cell suspension for 5 min at 450 × *g* to pellet the cells, discard the supernatant (see Note 11), and gently resuspend the cell pellet in 5 mL of proliferative medium.
10. Transfer the cell suspension into 25 cm^2 plastic flask: to obtain a sufficient amount of adhering myoblasts, and place at 37°C in an incubator, in a humidified 95% air/5% CO_2 atmosphere (see Note 12). Replace the medium with a fresh one every 3 days.
11. When at sub-confluence, myoblasts must be detached by mild trypsin treatment: remove the medium from the flask with a sterile pipette, rinse the flask with approximately 1 mL of trypsin-EDTA solution which must be quickly aspirated, add again 1 mL of trypsin-EDTA solution, and then incubate the flask at 37°C until the cells come off (see Note 13); add 2 mL of proliferative medium to block trypsin digestion, and split the cell suspension to three 25 cm^2 plastic flasks containing 4 mL each of pre-warmed proliferative medium.
12. It is suggested to store myoblasts samples at the second or third passage in culture. To do this, 6×10^5 to 1×10^6 detached cells must be placed in a 2.5 mL cryogenic vial containing 825 μL of proliferative medium plus 525 μL of FBS and 150 μL of DMSO. The vial containing the myoblasts must be pre-frozen overnight in isopropyl alcohol at −80°C, and finally transferred to liquid nitrogen for permanent storage.

3.2 Preparations of Myoblast and Myotube Samples on Glass Coverslips, for Light and Electron Microscopy

All the procedures should be carried out under sterile conditions.

1. *Proliferating myoblast samples*: detach myoblasts at 80% confluence by mild trypsinization (see Subheading 3.1, step 11 and Note 11); transfer the cell suspension into a sterile plastic tube, centrifuge for 5 min at 800 × *g*, remove the supernatant, and resuspend the cell pellet in 1 mL of proliferative medium; to estimate the concentration of living myoblasts, add 5 μL of cell suspension to 5 μL of trypan blue solution and count the Trypan-blue negative cells: at least 6×10^4 cells are to be planted onto a 22 × 22 mm sterile glass coverslip (see Note 14) in 35 mm petri dishes containing 1 mL of proliferative medium; let the cells adhere to the coverslip for 2 days in the incubator.
2. *Myotube samples*: for studying myoblast differentiation into myotubes, myoblast samples prepared as previously described (Subheading 3.2, step 1) are to be used; when myoblasts reach 80% confluence, replace the proliferative medium with 1 mL of the differentiation medium. After at least 7 days, myotubes start forming and the samples can be processed for microscopic analyses.

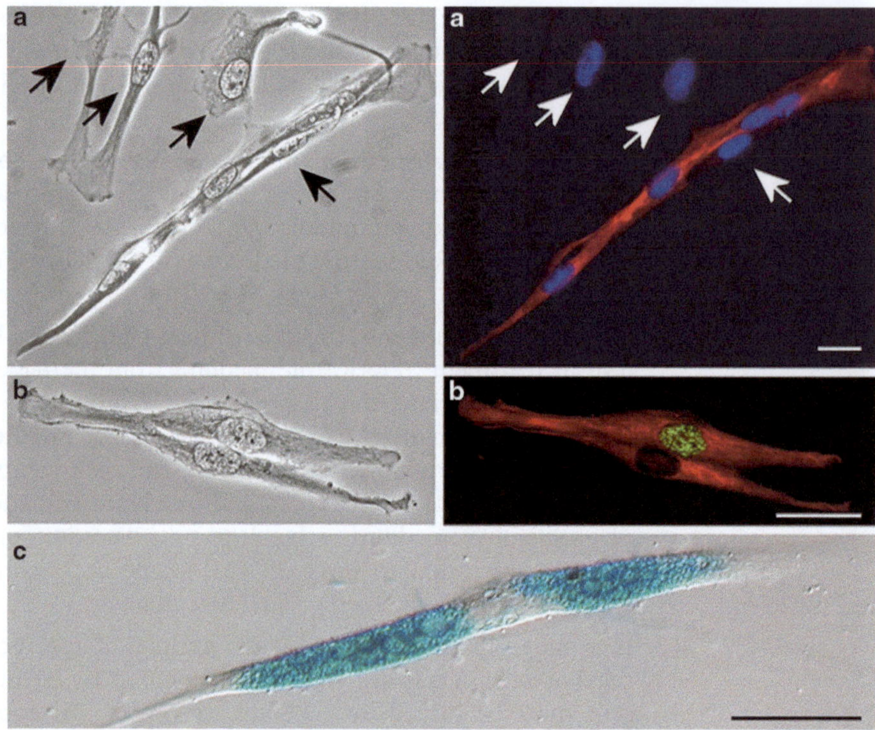

Fig. 1 Phase contrast (**a**) and fluorescence (**a′**) micrographs of a proliferating primary culture from isolated muscle cells: myoblasts are positive for desmin (*red* fluorescence) whereas fibroblasts (*arrows*) do not express this muscle-specific protein; nuclear DNA was stained with Hoechst 33258 (*blue* fluorescence). Phase contrast (**b**) and fluorescence (**b′**) micrographs of two desmin-positive myoblasts, one of which was in S-phase and did incorporate BrdU (*green* fluorescence). β-gal-positive myoblast (**c**) in a senescent culture. Bars: 30 μm

Myoblast and myotube preparations on glass coverslips are especially suitable for morphological and cytochemical procedures at light (bright-field, phase contrast and fluorescence) microscopy, as well as for cell monolayer embedding for transmission electron microscopy.

3.3 Light Microscopy: Detection of Proliferation and Senescence Markers

1. *Cytochemical detection of S-phase myoblasts* (Fig. 1a, b): to have a reliable index of myoblast proliferation, the S-phase fraction of cell cultures can be estimated by experiments of BrdU incorporation: incubate proliferating myoblasts for 2 h at 37°C with BrdU-containing proliferative medium; remove the glass coverslip by tweezers, and place it for fixation in a petri dish containing 70% ethanol at −20°C for 30 min (see Note 15). To estimate the percentage of S-phase myoblasts only (and exclude contaminating fibroblasts) the coverslip preparations are to be simultaneously immunolabeled for BrdU incorporation and for desmin. Briefly, after removing the ethanol and rehydrating with PBS for 10 min, incubate the myoblast coverslip in a

35 mm petri dish with 2 mL of 2N HCl for 15 min at RT to partially denature DNA, then with 0.1 M sodium tetraborate for 2 min to neutralize HCl, and wash with PBS for 5 min. Then, incubate the coverslip with the anti-desmin antibody for 1 h at RT, rinse with PBS for 5 min, incubate with Alexa 594-labeled secondary antibody for 1 h at RT, wash in PBS for 5 min, and finally incubate with the FITC-conjugated anti-BrdU antibody for 1 h at RT. After a 5 min washing with PBS, stain nuclear DNA with Hoechst 33258 for 5 min at RT, briefly rinse with PBS and finally mount upside-down with PBS/glycerol medium (see Note 16).

2. *Cytochemical detection of cell senescence* in vitro (Fig. 1c). Cultured cells are considered as senescent when more than 75% of the cells are positive for SA-β-Gal. To detect SA-β-Gal, remove the culture medium from the petri dish, wash in PBS, fix with 2% paraformaldehyde and 0.2% glutaraldehyde in PBS for 7 min at RT, remove the fixative solution and add the SA-β-Gal stain solution for 17 h at 37°C (no CO_2 control is necessary). Rinse with PBS and mount in PBS/glycerol medium as above (see Note 17).

3.4 Transmission Electron Microscopy: Processing for Morphology and Cytochemistry

All the procedures are carried out at RT, unless otherwise specified. Carefully make the solution changes in the petri dishes by gently using Pasteur pipettes in order to minimize cell detachment from the coverslip.

1. Remove the culture medium from the petri dish and put gently the fixative solution.
2. To process the samples for ultrastructural morphology, fix with 2% paraformaldehyde plus 2.5% glutaraldehyde in PBS for 2 h at 4°C, wash with PBS for 10 min and further fix with 1% OsO_4 solution for 1 h at 4°C, then rinse with PBS for 10 min. Dehydrate by using 60% acetone (5 changes of 5 min each) and then 100% acetone (5 changes of 5 min each), finally embed in EMbed-812 resin (3 changes of 15 min each at 60°C). This procedure allows optimal morphological preservation, but most cell molecules undergo denaturation, thus becoming barely detectable by cytochemical techniques (Figs. 2 and 3).
3. To process the samples for ultrastructural post-embedding cytochemistry, immunocytochemistry or in situ hybridization, fix with 4% paraformaldehyde in PBS for 2 h at 4°C, wash with PBS for 10 min, incubate with 0.5 M NH_4Cl solution for 30 min, and then rinse with PBS for 10 min. Dehydrate by using increasing concentrations (30, 50, 70, 90, and 100%) of ethanol (2 changes of 10 min each), finally embed in LR White resin (overnight at 4°C). This procedure is optimal for the

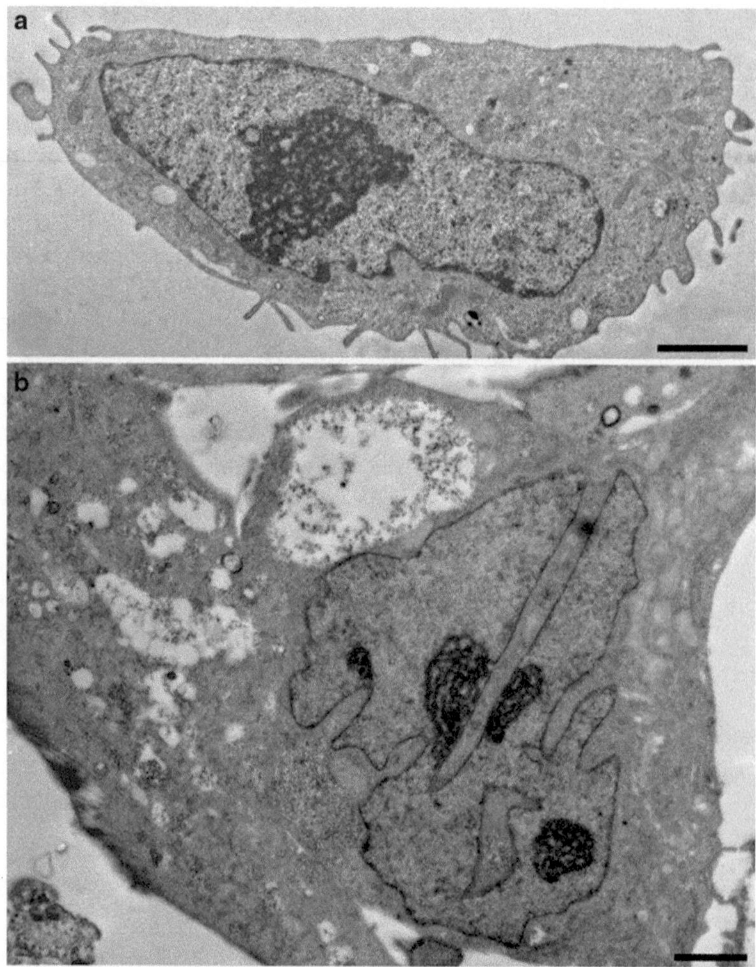

Fig. 2 Transmission electron micrographs of young (3rd culture passage) and senescing (15th passage) myoblasts processed for ultrastructural morphology (2% paraformaldehyde plus 2.5% glutaraldehyde fixation, 1% OsO_4 post-fixation, EMbed-812 resin embedding). Note the irregularly shaped nucleus and the cytoplasmic vacuolization typical of myoblast in vitro senescence. Bars: 2 μm

preservation of cell molecules, but the ultrastructural morphology is unclear, especially because of the lack of well defined cellular membranes (Fig. 4).

4. To obtain resin blocs, fill gelatin capsules with either EMbed-812 or LR White resin and place them upside-down over the coverslips. Put the samples in an oven for 24 h at 60°C. After polymerization, the coverslips can be detached from the resin blocs by dipping in liquid nitrogen for a few seconds: the cell monolayer will be visible on the resin bloc surface and will be ready to be sectioned (see Note 18).

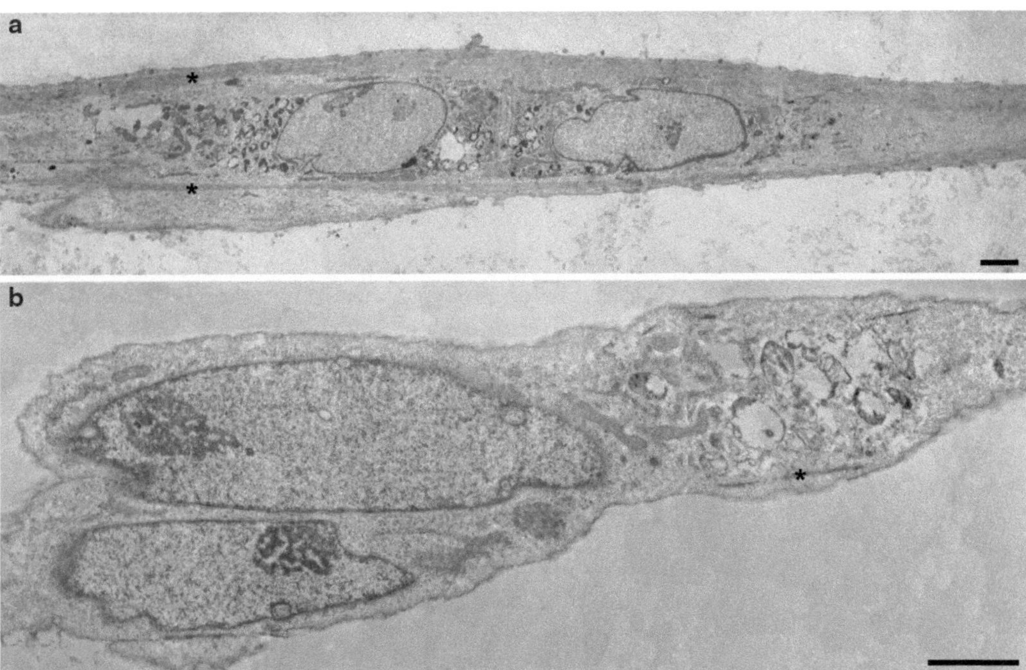

Fig. 3 Transmission electron micrographs of myotubes derived from young myoblasts (**a**) and from senescing myoblasts (**b**) processed for ultrastructural morphology (2% paraformaldehyde plus 2.5% glutaraldehyde fixation, 1% OsO_4 post-fixation, EMbed-812 resin embedding). The myotube in **a** shows an elongated shape, aligned nuclei and longitudinally arranged myofibrillar bundles (*stars*); the myotube in **b** shows an irregular shape, transversally aligned nuclei and scarce myofibrils (*star*). Bars: 2 μm

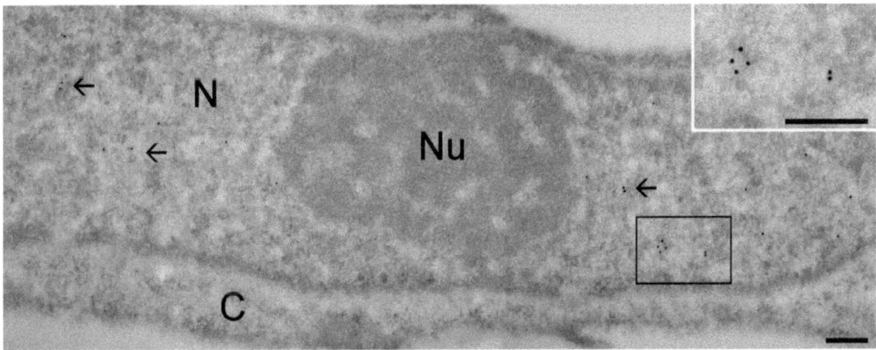

Fig. 4 Transmission electron micrographs of a myoblast processed for ultrastructural cytochemistry (4% paraformaldehyde fixation, LR White resin embedding). The ultrastructural morphology is less clear than in Fig. 2, but cytoplasm (C), nucleus (N), and nucleolus (Nu) are well recognizable. The ultrathin section has been submitted to a post-embedding immunocytochemical procedure using anti-polymerase II antibodies revealed by 12 nm colloidal gold grains (*arrows*) (this transcription factor drastically decreases in in vitro senescing myoblasts; Malatesta et al. (9)). The *inset* shows a high magnification detail of the labeling. Bars: 0.2 μm

4 Notes

1. Prepare the stock solutions at room temperature (RT) by stirring until the salts are completely dissolved.
2. Aliquots can be stored for several months at -20°C; avoid repeated freezing and thawing.
3. Paraformaldehyde dissolution is facilitated by heating; however, it is advisable to keep the temperature below 60°C, in order to avoid fixative degradation.
4. Cool the paraformaldehyde solution before adding glutaraldehyde, to avoid its thermic degradation.
5. Store at 4°C for a maximum of 2 weeks.
6. Aliquots of this staining solution may be store at -20°C in a dark bottle for a maximum of 6 months.
7. Osmium tetroxide crystals need 2–3 days to dissolve in water; therefore the stock solution must be prepared some days before use.
8. Due to the different density of the three components, it is advisable to add them in the cylinder in the reported order to facilitate mixing.
9. Although it is preferable to prepare fresh EMbed-812 resin each time, aliquots of the resin can be stored for a few weeks at -20°C.
10. The suspension must be gently shaken every 5 min.
11. Avoid aspirating the sediment.
12. To increase the myoblasts/fibroblasts ratio, plant the cells in a 25 cm^2 flask and place it in the incubator for 10 min to allow adhesion of fibroblasts which is much faster than myoblasts's adhesion. Then, remove the myoblast-enriched supernatant and transfer it in another flask.
13. To avoid excessive exposure to trypsin, monitor the progressive detachment of the cells under an inverted microscope; faster cell detachment may be promoted by gently rocking the flask.
14. For sterilization, wash the coverslips with 70% ethanol for at least 30 min, let them dry, and place overnight under UV light irradiation
15. 70% ethanol must be precooled at -20°C, and the coverslips must be dipped into the fixative, to avoid cell detachment: do not pour cold ethanol onto the coverslip surface. Fixed specimens can be stored in 70% ethanol at -20°C no more than 1 month.
16. These myoblast preparations on glass coverslips are suitable for cytochemical, immunocytochemical, and FISH procedures, and can be observed not only under bright field, phase contrast and

differential interference contrast, but also under fluorescence microscopy (which can hardly be used for cells grown on conventional plastic surfaces, which generally exhibit intense autofluorescence). As an example, some micrographs are shown in Figs. 1a–c.

17. It is not possible to perform the simultaneous labeling of cell cultures for desmin and SA-β-gal positivity: therefore, cell senescence can only be referred to the whole cell population (although myoblasts may be rather easily distinguished from fibroblasts by their thinner and more elongated shape).
18. The ultrathin sections from samples embedded in LR White resin can be used to detect a wide variety of molecular markers by using specific reagents (antibodies, enzymes, and nucleotidic probes); as an example, see Fig. 4.

Acknowledgments

Thanks are due to Mrs Paola Veneroni for her skilful technical assistance. Marzia Giagnacovo is a PhD student in receipt of a fellowship from the Dottorato di Ricerca in Biologia Cellulare (University of Pavia).

References

1. Tews DS, Goebel HH (2005) Diagnostic immunohistochemistry in neuromuscular disorders. Histopathology 46:1–23
2. Vogel H, Zamecnik J (2005) Diagnostic immunohistology of muscle diseases. J Neuropathol Exp Neurol 64:181–193
3. Malatesta M, Perdoni F, Muller S, Zancanaro C, Pellicciari C (2009) Nuclei of aged myofibres undergo structural and functional changes suggesting impairment in RNA processing. Eur J Histochem 53(97–106):e12
4. Malatesta M, Meola G (2010) Structural and functional alterations of the cell nucleus in skeletal muscle wasting: the evidence in situ. Eur J Histochem 54:e44
5. Hayat MA (1989) Principles and techniques of electron microscopy, 3rd edn. CRC, Boca Raton, FL
6. Cardani R, Baldassa S, Botta A, Rinaldi F, Novelli G, Mancinelli E et al (2009) Ribonuclear inclusions and MBNL1 nuclear sequestration do not affect myoblast differentiation but alter gene splicing in myotonic dystrophy type 2. Neuromuscul Disord 19:335–343
7. Perdoni F, Malatesta M, Cardani R, Giagnacovo M, Mancinelli E, Meola G et al (2009) RNA/MBNL1-containing foci in myoblast nuclei from patients affected by myotonic dystrophy type 2: an immunocytochemical study. Eur J Histochem 53: 151–158
8. Giagnacovo M, Costanzo M, Cardani R, Veneroni P, Pellicciari C, Meola G (2011) Ultrastructural features of myotubes derived from myoblasts of patients affected by myotonic dystrophy type 2, after senescence in vitro. Eur J Histochem 55(Suppl 1):15
9. Malatesta M, Giagnacovo M, Renna LV, Cardani R, Meola G, Pellicciari C (2011) Cultured myoblasts from patients affected by myotonic dystrophy type 2 exhibit senescence-related features: ultrastructural evidence. Eur J Histochem 55:e26
10. Dimri GP, Lee X, Basile G, Acosta M, Scott G, Roskelley C et al (1995) A biomarker that identifies senescent human cells in culture and in aging skin in vivo. Proc Natl Acad Sci USA 92:9363–9367

Chapter 7

Cardiac Stem Cell Senescence

Daniela Cesselli, Federica D'Aurizio, Patrizia Marcon, Natascha Bergamin, Carlo Alberto Beltrami, and Antonio Paolo Beltrami

Abstract

Cellular senescence processes affecting tissue resident stem cells are considered, at present, an hallmark of both aging and age-related pathologies. Therefore it is mandatory to address this problem with adequate techniques that could highlight the molecular alterations associated with this complex cellular response to stressors. Here we describe methods to characterize cardiac stem cell (CSC) senescence from a molecular and functional standpoint.

Keywords Cardiac stem cells, Cellular senescence, Telomere dysfunction, Persistent DNA-Damage Response, p16, Telomere length, Cell migration, Cell differentiation

1 Introduction

Cellular senescence is a complex cellular response to a variety of stressors that results in a permanent arrest in cell proliferation (1–5), thus representing a powerful anti-oncogenic mechanism (6, 7). However, it has been demonstrated that cells that express senescence markers accumulate at sites of chronic age-related pathology, such as osteoarthritis, atherosclerosis, and chronic heart failure (6, 8–16), linking cell senescence with aging and age-related diseases in vivo (17). Moreover, accumulated evidences indicate that cell senescence can affect the stem cell compartment and this is associated with a reduction in the stem cell function (2, 6, 18). Therefore, following the evolutionary theory of antagonistic pleiotropy, stem cell senescence can be considered a double edged-sword that exerts both a tumor-suppressor effect, by preventing the expansion of injured self-renewing cells, and detrimental effects, contributing to tumor invasiveness in a paracrine fashion or to aging by causing stem cell arrest or attrition (cancer-aging hypothesis) (6, 7).

Despite its involvement in a variety of physiological and pathological alterations, the lack of cellular markers specific enough to identify it in tissues, has hampered research aimed at identifying

Kursad Turksen (ed.), *Stem Cells and Aging: Methods and Protocols*, Methods in Molecular Biology, vol. 976, DOI 10.1007/978-1-62703-317-6_7, © Springer Science+Business Media, LLC 2013

the relevance of cellular senescence in vivo (19). Populations of cells, both in vitro and in vivo, are heterogeneous, being composed of variable fractions of both proliferating and senescent cells. The mechanism responsible for this feature is still debated (1). The classical view attributes it to a vicious cycle between generation of Reactive Oxygen Species (ROS) and damage to the DNA, and more in general, to macromolecules, while more recent views consider increased ROS generation a typical feature of a senescence response, rather than a cause (20). In any case, DNA-damage-induced cellular senescence is associated with a persistent DNA damage response (DDR) (4, 21). The recent demonstration that telomeric DNA, if damaged, are irreparable and trigger persistent DDR and cellular senescence (22), stressed the importance of dysfunctional telomeres (23).

The approach here described is a cross-sectional study of populations of Cardiac Stem Cells at low Population Doubling (PD) level to verify if, even in populations of young cells, donor age and pathological state were reflected in an increased prevalence of senescent and dysfunctional cells. Therefore, the purpose of the present protocol is to provide an experimental platform to evaluate, in vitro, both intrinsic cell senescence and functional impairment of CSC. Figure 1 depicts the main strategy adopted and the approximate number of cells required for each assay, considering plastic ware and instrumentation commonly available in many research laboratories. However, it is possible to reduce the number of assayed cells and automate some of the analysis (see Note 1).

2 Materials

Prepare every buffer that will be in contact with living cells by employing non-pyrogenic, ultrapure water for cell cultures. Use disposable cell culture plasticware, where possible, or wash glassware with detergents that leave minimal chemical contamination (7×, MP Biomedicals, USA). Sterilize cell culture glassware using a hot air oven at 190°C. Square brackets indicate final concentration. Sterilize every solution that comes in contact with living cells under a laminar flow hood by employing disposable sterile filter systems (0.22 μm pore size).

2.1 Cell Dissociation

1. Sterile forceps, scissors, scalpels, and petri dishes.
2. Basic Dissociation Buffer: Reconstitute Minimum Essential Medium Joklik (J-MEM) in 1 L ultrapure H_2O, add 0.7 g HEPES (3 mM), 0.3 g Glutamine (2 mM), Insulin (20 U/L), Penicillin (100 U), and Streptomycin (100 μg/L). pH to 7.3 with NaOH. Filter under hood the solution with a luerlock syringe filter (0.22 μm pore size).
3. Incubation Buffer: add BSA to Basic Buffer (0.5% w/v), adjust the pH to 7.3.

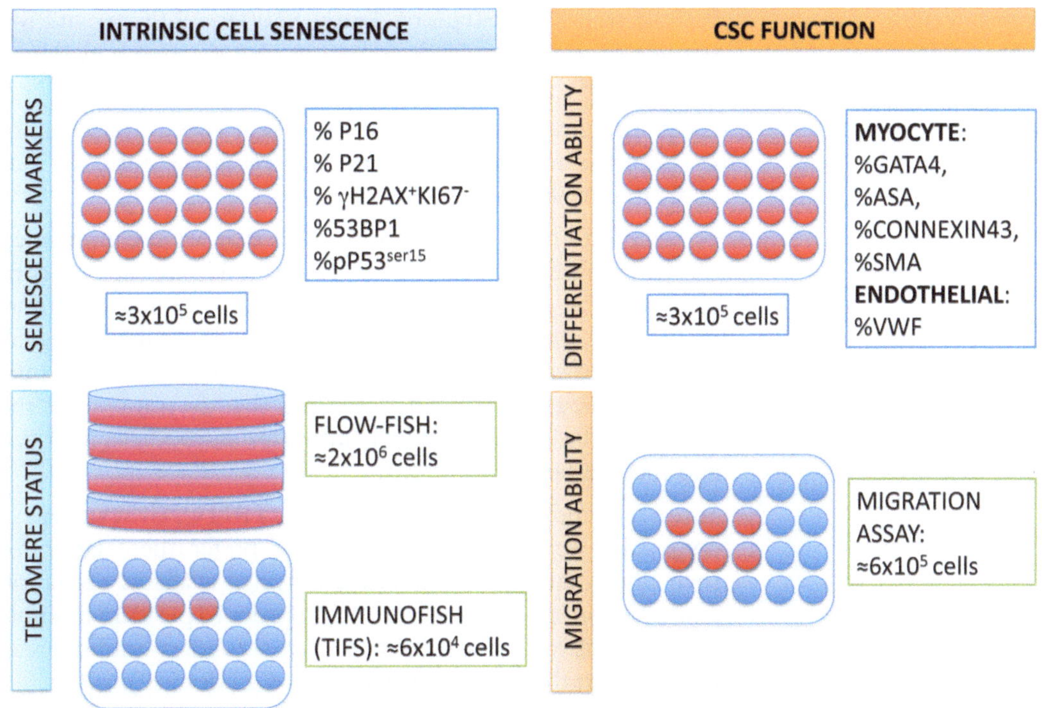

Fig. 1 Schematic representation of the protocols employed to assess of level of CSC senescence

4. Collagenase solution: add type II Collagenase to Basic Buffer (≈50 CDU/mL). Filter under hood the solution with a luer-lock syringe filter (0.22 μm pore size).
5. Calcium free, Magnesium free Hank's Balanced Salt Solution (HBSS): add the content of 1 package (Sigma-Aldrich, USA) in 1 L ultrapure H_2O, adjust the pH to 7.3. Sterilize by filtration.
6. Dissociation solution: ready to use trypsin EDTA solution (0.5 g/L porcine trypsin and 0.2 g/L EDTA·4Na in Hank's Balanced Salt Solution with phenol red).

2.2 Immuno-fluorescence

1. Sterile coverslips: sterilize round glass coverslips (8 mm diameter) by placing them in a glass petri dish, using a hot air oven at 190°C.
2. 4% Paraformaldehyde: weight 4 g of paraformaldehyde. Under a fume hood, add it to a becker containing 90 mL of pre-warmed (≈60°C) bidistilled H_2O, placed on a heated stir plate. Keep heating and stirring paraformaldehyde for at least 1 h. Add few drops of 1N NaOH to clear the solution. Cool down the solution, then add 10 mL of 10× PBS and bring to 100 mL with bidistilled H_2O, if necessary. Check the pH and adjust it at 7.4. Keep it refrigerated, and use it within 1 week (see Note 2).
3. Phosphate-buffered saline (PBS).

4. 0.1% Triton X: add 50 μL of Triton X100 to 49.95 mL of PBS, stir gently until dissolved, and store at 4°C. Use it within 1 week.
5. Blocking buffer (10% serum in PBS): add 100 μL of serum of the same animal species where the secondary antibody was generated to 900 μL of PBS.
6. Primary antibodies: employ antibodies that are recommended for the application of choice (see Note 3). Titrate the best concentration by employing positive controls (cell lines that normally do express the antigen of choice or pharmacological treatment of cells in order to induce the expression of the antigen), and negative controls (cell lines that do not express the antigen to be tested). Table 1 illustrates antibodies employed and their respective dilutions.
7. Secondary antibodies: we tested FITC, TRITC, Cy5, and Alexa conjugated secondary antibodies. We prefer these latter for their photostability and the wide range of available excitation and emission spectra (see Note 4). In order to perform multiple labeling, it is recommended to employ secondary antibodies all generated in the same animal species (e.g., donkey).

2.3 Flow-FISH

There are kits (e.g., Telomere PNA Kit/FITC for Flow Cytometry; Dako, Denmark) that provide most of the reagents in an easy to use format. Alternatively, very well-written and accurate protocols are available in the literature (24). To describe in detail the latter is beyond the scope of this article.

1. Hybridization solution: it is provided as a ready to use reagents in the Dako kit.
2. Telomere probe: this is a fluorescein-conjugated peptide nucleic acid (PNA) probe that recognizes telomeres from all vertebrate nucleated cells. It is provided in the kit in hybridization solution.
3. Wash solution: it is provided by the kit in a 10× format and needs to be diluted, prior to use, to its working concentration with bidistilled H_2O.
4. DNA staining solution: consists of a 10× stock of Propidium Iodide and RNase A, which has also to be diluted prior to use in bidistilled H_2O.
5. Control cells: several cells have been employed as controls for Flow-FISH, among these the 1301 human tetraploid T cell leukemia cell line and freshly isolated bovine
6. Thymocytes.
7. Expansion medium for the 1301 cell line. Add to 800 mL of RPMI 1640 culture medium the following additives: 10 mL of 100× glutamine stock solution (2 mM), 100 mL of FBS (10%). Bring to 1 L with RPMI 1640, sterilize by filtration.

Table 1
Antibodies and summary of the immunofluorescence protocols employed

Cell staining						
		Primary antibody			Secondary antibody[b]	
Staining	Unmasking treatment[a]		Dilution	Incubation	Fluorochrome	Dilution
OCT4	TRITONX-100 0.1%	ABCAM, RP	1:150	O/N, 4°C	A488	1:800
NANOG	TRITONX-100 0.1%	ABCAM,RP	1:150	O/N, 4°C	A488	1:800
SOX2	TRITONX-100 0.1%	MILLIPORE, RP	1:200	O/N, 4°C	A488	1:800
c-Kit	/ /	R&D, GP DAKO, RP	1:100 1:100	2 h, 37°C	A488/A555 A488/A555	1:800 1:800
$p16^{INK4A}$	TRITONX-100 0.1%	MTM, MM	Prediluted	O/N, 4°C	A488	1:800
p21	TRITONX-100 0.1%	CALBIOCHEM, MM	1:40	O/N, 4°C	A488	1:800
Phospho-$p53^{ser15}$	TRITONX-100 0.1%	CELL SIGNALING, MM	1:500	O/N, 4°C	A488	1:800
Ki67/γ H2AX[c]	TRITONX-100 0.1%	LEICA, RP MILLIPORE, MM	1:1,000 1:500	O/N, 4°C 2 h, 37°C	A555 A488	1:600 1:600
53BP1	TRITONX-100 0.1%	CELL SIGNALING, RP	1:100	O/N, 4°C	A555	1:800
GATA4	TRITONX-100 0.1%	S. CRUZ, RP	1:40	O/N, 4°C	A488	1:800
vWF	/	SIGMA, RP	1:40	2 h, 37°C	A488	1:800
CD31	/	DAKO, MM	1:50	2 h, 37°C	A488	1:800
SMA	TRITONX-100 0.1%	DAKO, MM	1:50	2 h, 37°C	A488	1:800
CONNEXIN 43/α-SA	TRITONX-100 0.1%	S. CRUZ, RP SIGMA, MM	1:50 1:200	2 h, 37°C 2 h, 37°C	A488 A555	1:800 1:800

RP rabbit polyclonal, *GP* goat polyclonal, *MM* mouse monoclonal, *vWF* von Willebrand factor, *SMA* smooth muscle actin, *α-SA* α-sarcomeric actin, *O/N* overnight, *h* hours, *RT* room temperature

[a]Incubation: 10 min, RT

[b]Incubation: 1 h 37°C

[c]Staining will proceed with this order: rabbit anti-Ki-67, A488 donkey anti-rabbit, mouse anti-γH2AX, A555 donkey anti-mouse

2.4 Immuno-FISH

In this case too, we recommend the use of kits (e.g., Telomere PNA FISH Kit/FITC; Dako, Denmark) that provide both most of the reagents and the PNA probes.

1. Primary antibody: 53BP1 (Cell Signaling, rabbit polyclonal), diluted 1:100.
2. Secondary antibody: we employed an Alexa 555 labeled donkey anti-rabbit antibody, given the good thermal stability of Alexa dyes. Working dilution was 1:800 in PBS.
3. Tris-Buffered Saline, pH 7.5 (TBS): dissolve 1.4 g of Tris-Base (11.5 mM), 6.0 g of Tris–HCl (38.1 mM), and 8.75 g of NaCl (136.9 mM), bring to 1 L with bidistilled H_2O, adjust pH to 7.5.
4. 3.7% buffered formaldehyde: add 10 mL of the 37% buffered formaldehyde stock to 90 mL of TBS.
5. Prepare three coplin jars containing cold 70, 85, and 96% ethanol.
6. Telomere PNA probe: same probe as above, provided in hybridization solution.
7. Rinse solution and wash solution: provided with the kit as 50× concentrate to be diluted in bidistilled H_2O.

2.5 Functional Assays

In both functional assays, we tested culture expanded CSC at the third passage in vitro (P3).

2.5.1 Differentiation Assay

1. Muscle cell differentiation medium: Reconstitute low glucose DMEM in 1 L of ultrapure H_2O_2, adjust the pH to 7.3, sterilize by filtration; reconstitute MCDB-201 in 1 L ultrapure H_2O, adjust the pH to 7.3, sterilize by filtration. Prepare 1 L of Basal Medium mixing 600 mL of low glucose DMEM with 400 mL of IMCDB-201. Add to 900 mL basal medium the following additives: 1 g linoleic acid-BSA (1 mg/mL), 3.92 μg dexamethasone (10^{-8} M), 28.9 mg ascorbic acid-2 phosphate (10^{-4} M), 5 mg insulin (5 μg/mL), 5 mg transferrin (5 μg/mL), 5 μg sodium selenite (30 nM), 50 mL fetal bovine serum (FBS) (5%), 10 μg basic Fibroblast Growth Factor (bFGF) (10 ng/mL), 10 μg Vascular Endothelial Growth Factor (VEGF) (10 ng/mL), and 10 μg Insulin-like Growth Factor-1 (IGF1) (10 ng/mL). Bring to 1 L with basal medium and sterilize by filtration.
2. Endothelial cell differentiation medium: add to the 500 mL bottle of Endothelial Cell Basal Medium-2 (EBM-2, Lonza, Basel, Switzerland), the following supplements (BulletKit, Lonza, Basel, Switzerland): 0.5 mL human Epidermal Growth Factor, 2.0 mL bFGF, 0.5 mL VEGF, 0.5 mL Ascorbic Acid, 0.2 mL Hydrocortisone, 0.5 mL Long R3-IGF-1, 0.5 mL Heparin, and 10 mL FBS.

2.5.2 Migration Assay

1. 24-Well Cell Migration chamber (CultrexTrevigen). It consists of a cell culture insert, that is a simplified Boyden chamber with an 8 μm polyethylene terephthalate (PET) membrane.
2. Serum free medium: basal medium (BM) described above.
3. Chemoattractant: add 10 mL of FBS to 90 mL of BM and sterilize by filtration (10% v/v).
4. Quenching medium: add 5 g of Bovine Serum Albumin (BSA) to 100 mL BM (5% w/v), adjust the pH to 7.3 and sterilize by filtration.
5. Vital cell labeling: dilute 50 μg Calcein-AM in 25 μL sterile DMSO to obtain a 2 mM stock solution (that can be stored at -20°C).
6. Washing solution: Hank's Balanced Salt Solution (HBSS) without phenol red.
7. Dissociation solution: ready to use trypsin EDTA solution (0.5 g/L porcine trypsin and 0.2 g/L EDTA•4Na in HBSS without phenol red).

3 Methods

3.1 Cell Isolation and In Vitro Expansion of CSC

1. Collect aseptically a cardiac sample into a 50 mL conical sterile tube containing a volume of Basic Dissociation Buffer (BDB) equal to the volume of the sample.
2. Warm up at Room Temperature for about 30 min BDB, Incubation Buffer (IB), and Collagenase solution.
3. In a petri dish containing ≈5 mL of BDB, remove pericardial adipose tissue and macroscopically apparent areas of fibrosis.
4. Transfer the sample to a new petri dish containing ≈3 mL of BDB solution. Start mincing the fragment with a scalpel, then with scissors until the fragments have a dimension not larger than 1 mm^3.
5. Add about 7 mL of BDB to the petri dish and collect medium and fragments in a 15 mL conical sterile tube. Let the fragments sediment and discard the supernatant.
6. Resuspend the fragments in other 10 mL BDB and centrifuge it at 500 ×g for 1 min. Discard the supernatant.
7. Add to the pellet a volume of collagenase solution of at least two times the volume of the fragments.
8. Incubate the suspension at 37°C for about 20 min in a tube rotator (fragments obtained from normal donors are incubated for a shorter period, while those collected from failing hearts require a longer incubation period). If large undigested fragments are still present, incubate for an extra 5 min and verify again.

9. At the end of the incubation period, gently shake the bottom of the Falcon Tube and add 5 mL of IB and gently pipette it for 4–5 min.
10. Centrifuge cell suspension at $50\times g$ for 1 min to remove cardiac myocytes. Collect the supernatant.
11. Pre-wet a 40 μm Mesh Filter by filtering 5 mL of sterile, fresh HBSS using one 5 mL sterile serological pipette into a 50 mL Falcon Tube.
12. Filter the cell suspension through the pre-wet 40 μm Mesh Filter.
13. Centrifuge the filtered suspension at $500\times g$ for 5 min and resuspend them into 1 mL of proliferation medium (18).
14. Count the cells using a Burker Counting Chamber.
15. Seed 1.5×10^6 freshly isolated cells into 100 mm-dishes.
16. Subcultures: CSC in primary culture grow forming discrete colonies. It is important to avoid that colonies become over-confluent. We do not expect that cells in primary culture will cover the entire plate. To detach cells from substrate, plates are washed two times by adding and completely removing 5 mL of HBSS. Subsequently, 3 mL of Trypsina-EDTA solution is added to the dish and incubated at room-temperature (RT) for not more than 10 min. As soon as cells detach, 7 mL of culture medium is added to the plate to block the enzymatic digestion. Further passages will be performed keeping constant seeding density (2×10^3 cells/cm^2) and population doublings (3, 4) before further passaging. Culture medium is replaced every 3–4 days.

3.2 Immunofluorescence

1. Use sterile forceps to dispense sterile coverslips into the single wells of a 24-well multiwell plate.
2. Detach cells from a 70% confluent petri dish, as if for subculturing, and resuspend them in growth medium.
3. Seed cells at a concentration of 1×10^4/cm^2 on top of the fibronectin-coated coverslips, bring culture medium to 500 μL/well and incubate the cells at 37°C, 5% CO_2.
4. Fix the cells 1 day after seeding. To do so, first remove culture medium from the wells using a transfer pipette, then wash two times the cells by adding and removing ≈1 mL PBS per well. Last, add ≈0.5 mL of 4% paraformaldehyde for 15 min at room temperature (see Note 5). After fixation, wash the cells three times by adding and removing ≈1 mL PBS per well. Change PBS every 5 min. Fixed cells may be kept for not more than 3 days at +4°C into PBS filled wells of multiwell plates. Seal plate covers with parafilm.
5. In order to detect intracellular antigens, cells must be permeabilized with 0.1% Triton X for 10 min at room temperature.

Remove the excess of Triton X washing the cells three times by adding and removing ≈1 mL PBS per well. Change PBS every 5 min.

6. In order to saturate unspecific binding sites, incubate slides with blocking buffer for 30 min at room temperature.
7. Dilute the primary antibodies in PBS to their working concentration (Table 1).
8. Place a clean piece of parafilm in a moist chamber containing a wet paper towel. Dispense, on the parafilm, 30 μL of diluted primary antibody in a single drop. Label with a marker the parafilm just below the drop to indicate the target antigen. Carefully remove the coverslip from the well by employing a pair of dissecting forceps with sharp tips. First, remove the excess of blocking buffer placing the edge of the slide on a piece of blotting paper, then lay the face of the slide containing the cells on the drop containing the primary antibody. See Table 1 for the incubation conditions.
9. After incubation, detach the coverslips from parafilm and place them (with cells on the upper side) each into a well of a 24-well multiwell plate filled with PBS. Wash the cells five times by adding and removing ≈1 mL PBS per well. Change PBS every 2 min.
10. Place a clean piece of parafilm in a moist chamber containing a wet paper towel. Dispense 30 μL of diluted secondary antibody in a single drop. Label with a marker the parafilm just below the drop to indicate the target antigen. Carefully remove the coverslip from the well by employing a pair of dissecting forceps with sharp tips. Lay the face of the slide containing the cells on the drop containing the secondary antibody. Incubate the cells for 1 h at 37°C.
11. After incubation, detach the coverslips from parafilm and place them (with cells on the upper side) each into a well of a 24-well multiwell plate filled with PBS. Wash the cells five times by adding and removing ≈1 mL PBS per well. Change PBS every 2 min. Wash for 2 additional minutes by adding 1 mL of distilled water.
12. Add 10 μL of Vectashield containing DAPI (Vector, USA)—or a similar anti-fading anti-bleaching mounting medium for fluorescence—on the top side of a glass slide. Label the slide with a permanent marker indicating the name of the cell line, target antigen name and fluorochrome employed.
13. Carefully remove the coverslip from the well by employing a pair of dissecting forceps with sharp tips. Lay the face of the slide containing the cells on the drop. With a clean paper towel delicately remove the excess of PBS and Vectashield that possibly emerge from the sides of the coverslip (see Note 6). Seal the edges of the coverslip to the slide with a small drop of nail polish.

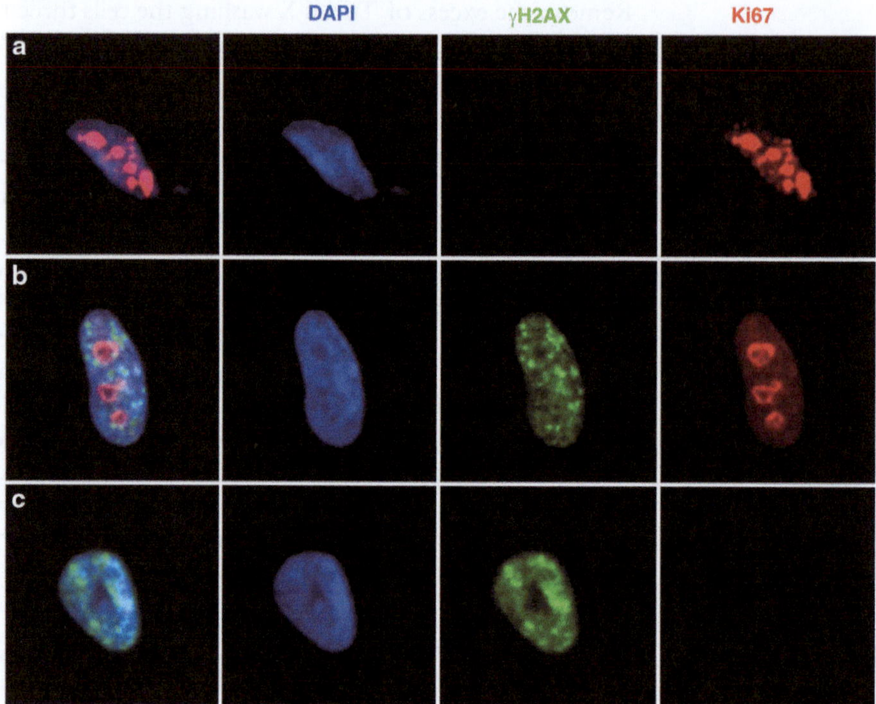

Fig. 2 Evaluation of the cells with a persistent DNA-damage response (DDR). **a**, **b**, and **c** represents three cells stained by Ki67 (*red* fluorescence), γH2AX (*green* fluorescence), and DAPI (*blue* fluorescence). Only cell C was counted as cell with a persistent DDR since positive for γH2AX and negative for Ki67

14. *Analysis*: samples are analyzed employing an epi-fluorescence microscope equipped with excitation and emission filters appropriate for the acquisition of the fluorochromes used in the staining. Samples are usually analyzed using a 40× objective. At least 400 cells, identified by the blue fluorescence of DAPI staining, are scored as positive or negative for the assayed antigens. Specifically, it will be determined the fraction of cells positive for p16, p21, phosphoP53^{ser15}, and Ki67. The fraction of cells characterized by a permanent DNA-Damage Response (DDR) is identified counting the cells positive for γH2AX that do not express Ki67 (Fig. 2).

3.3 Flow-FISH

Procedure is illustrated in Fig. 3a.

1. Expand in vitro the 1301 cell line in the appropriate culture medium at 37°C, 5% CO_2. Keep the cell density between 3 and 9×10^5 cells/mL.
2. Prior to starting the experiment, pre-warm a heating block to 82°C.
3. Label four 1.7 mL centrifuge tubes as "A," "B," "C," and "D."

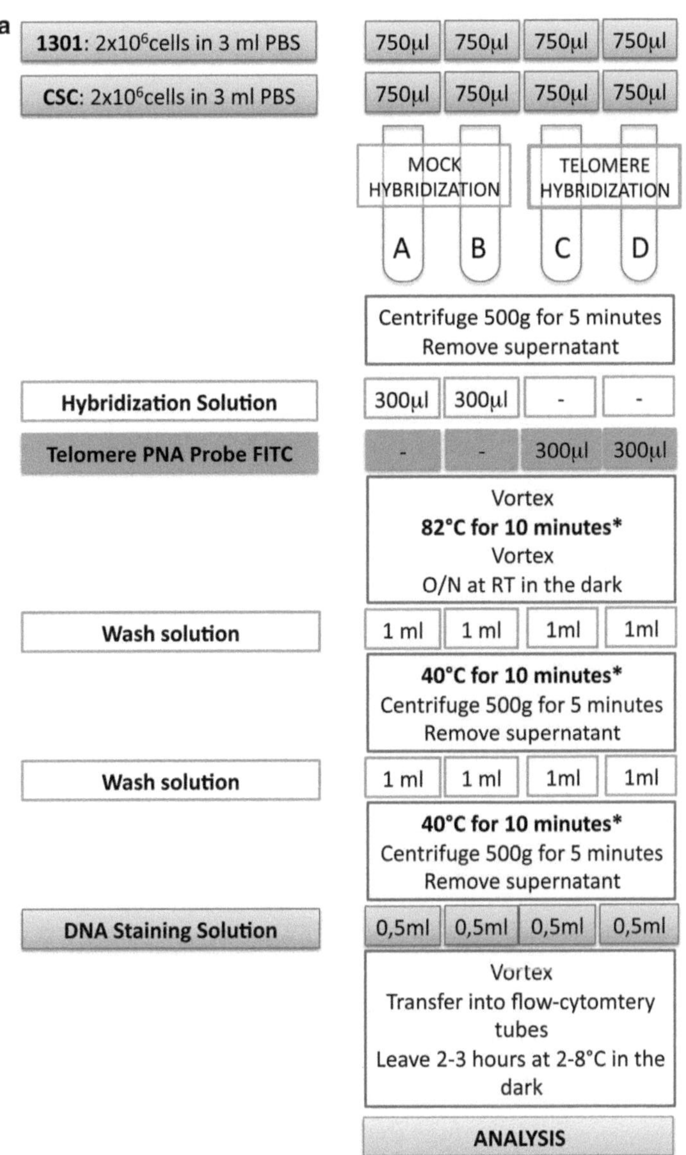

Fig. 3 Flow-FISH. (**a**) Protocol. (**b–g**) Flow-FISH analysis: gating strategy. Mock-hybridized (**b–d**) and Telomere Probe-hybridized (**e–g**) cells were identified on the basis of their physical properties (**b**, **e**). Cell doublets were excluded from the analysis by evaluating pulse with vs. peak value of FSC (**c**, **f**). CSC (*red* gate) and 1301 (*blue* gate) cells were distinguished on the basis of their DNA content (**d**, **g**). Mean fluorescence intensity of gated cells were employed to compute relative telomeric length (see text)

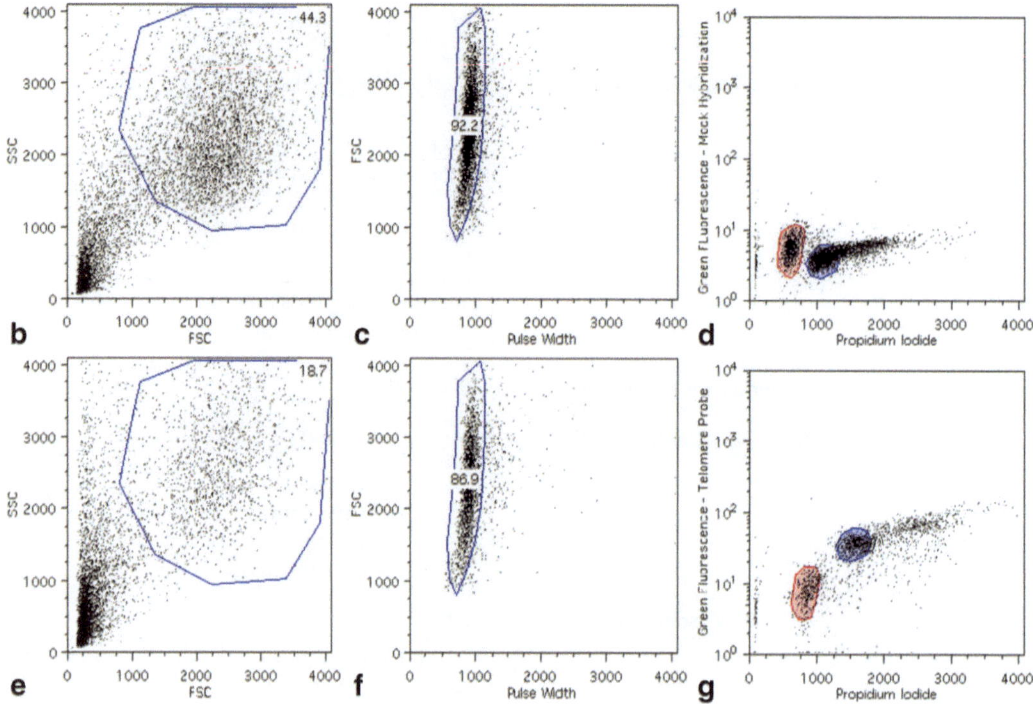

Fig. 3 (continued)

4. To carry out the experiment, collect 2×10^6 cells, resuspend them in 2 mL PBS and dispense 500 μL of cell suspension in the four 1.7 mL centrifuge tubes, labeled "A," "B," "C," and "D."
5. Detach CSC from a 70% confluent petri dish, as if for subculturing, and resuspend them in PBS. Collect 2×10^6 cells in 2 mL of PBS and dispense 0.5×10^6 cells (500 μL of cell suspension) to the previously mentioned "A," "B," "C," and "D" tubes.
6. Add 500 μL of PBS in each tube.
7. Centrifuge the vials at $500 \times g$ for 5 min and remove the supernatant as much as possible.
8. Add 300 μL of Hybridization Solution to tubes "A" and "B" and 300 μL of Telomere PNA Probe/FITC in hybridization solution to the remaining vials. Place the tubes in the heating block set at 82°C for 10 min.
9. Remove the vials from the block, mix by vortex and incubate overnight at room temperature in the dark.
10. The following day, pre-warm the heat block at 40°C. Add 1 mL wash solution to each vial, vortex, and place the tubes in the heat block for 10 min.
11. Vortex again the cells and pellet them by centrifugation ($500 \times g$ for 5 min). Discard the supernatant into a waste container for formamide.

12. Repeat once the washing steps. Discard the supernatant.
13. Add 500 μL DNA staining solution to each tube and vortex. Transfer the cells to cytometer tubes labeled as the corresponding 1.7 mL ones. Incubate the cells at 4°C for not less than 3 h.
14. *Analysis*:
    (a) Analyze the samples with a Flow-Cytometer equipped with a 488 laser, acquiring DNA labeling (FL3 peak height fluorescence -FL3-H-) in a linear mode and FITC peak height fluorescence (FL1-H) on a logarithmic scale.
    (b) Gating strategy, as shown in Fig. 3b, is meant to exclude cells in S and G_2/M phases of the cell cycle from the analysis. Mean green fluorescence intensity of both CSC and 1301 cells should be collected. Furthermore, mock hybridization samples (vials A and B) will be compared with telomere hybridization samples (vials C and D).
    (c) DNA index of both CSC and control cells should be computed in a separate experiment, according to well-established methods (25) or taking advantage of commercially available kits.
    (d) Relative Telomeric Length of the sample cells compared to the control cells will be computed in the following way: (mean green fluorescence of CSC with probe – mean green fluorescence of CSC without probe) × DNA index of 1301 cells × 100/(mean green fluorescence of 1301 cells with probe – mean green fluorescence of 1301 cells without probe) × DNA index of CSC.

3.4 Immuno-FISH

1. This procedure consists of an immunofluorescence step, which is performed as in Subheading 3.2 and employs a 53BP1 primary antibody and an Alexa 555 labeled secondary antibody. Do not mount the coverslips at the end of the immunostaining, but proceed with the following FISH protocol. Coverslip manipulations are carried out by employing the same dissecting forceps employed in Subheading 3.2.
2. Day 1:
    (a) Use a 24-well multiwell plate to perform the pretreatment. Dispense in distinct wells: 1 mL of TBS, 1 mL of 3.7% formaldehyde, and 1 mL of TBS. Immerse the coverslip in the first well containing TBS for about 2 min, then move it to the well containing formaldehyde and incubate it at room temperature for 10 min. Remove the coverslip from the fixative and move it to the adjacent TBS containing well. Change TBS five times, every 2 min.
    (b) Immerse the coverslips in the coplin jars containing ethanols. Hold the coverslips with forceps and keep them in each ethanol for 2 min. After the last ethanol, air dry the cells.

(c) Add 10 μL of Telomere PNA Probe/FITC on the top side of a glass slide. Place the slides in a preheated incubator set at 80°C for 5 min (see Note 7).

(d) Incubate the slides over/night at room temperature (see Note 8).

3. Day 2:

(a) Pre-warm ≈50 mL wash solution in a coplin jar immersed into a water bath set at 65°C.

(b) Add about 7 mL of rinse solution to a 100 mm diameter petri dish. Immerse the slide in the dish and gently remove the coverslip from the slide.

(c) Hold the slide for 5 min in hot (65°C) wash solution.

(d) Immerse the coverslips in the coplin jars containing ethanols. Hold the coverslips with forceps and keep them in each ethanol for 2 min. After the last ethanol, air-dry the cells.

(e) Add 10 μL of Vectashield containing DAPI on the topside of a glass slide. Label the slide with a permanent marker indicating the name of the cell line, Telomeres FITC and 53BP1 Alexa 555.

(f) Carefully remove the coverslip from the well by employing a pair of dissecting forceps with sharp tips. Lay the face of the slide containing the cells on the drop. With a clean paper towel delicately remove the excess of PBS and Vectashield that possibly emerge from the sides of the coverslip (see Note 6). Seal the edges of the coverslip to the slide with a small drop of nail polish.

4. *Analysis: Telomere dysfunction Induced Foci* (TIF) are evaluated by employing an epifluorescence microscope equipped with a oil immersion 63× or 100× objective and a camera for the image acquisition. A cell has to be considered as positive for TIF when the number of 53BP1 colocalizing with the telomere hybridization spots is ≥50% of the total 53 BP1 spots. At least 100 cells per sample have to be evaluated (23).

3.5 Functional Assays

3.5.1 Differentiation Assay

1. Use sterile forceps to dispense sterile coverslips into the single wells of a 24-well multiwell plate.
2. Detach cells from a 70% confluent petri dish, as if for subculturing, and resuspend them in appropriate (muscle or endothelial cell) differentiation medium.
3. Seed cells at a concentration of $0.5–1 \times 10^4/\text{cm}^2$ on top of the coverslips, bring culture medium to 500 μL/well and incubate the cells at 37°C, 5% CO_2.
4. Allow cells to become confluent and culture cells for either 4 weeks (muscle differentiation) or 2 weeks (endothelial cell differentiation). Change the medium twice a week.
5. At the end of the protocol, fix the cells as in Subheading 3.2.

3.5.2 Migration Assay

1. The day prior to assay, starve the cells in serum-free medium.
2. 24 h after starvation, detach cells from a 70% confluent petri dish, as if for subculturing, and resuspend them in BM.
3. Count the cells and adjust concentration to 10^6 cells/mL.
4. Prepare a standard curve as follows:
   (a) Dispense, in triplicate, a serial dilution series (e.g., 100,000, 50,000, 25,000, 12,500, 6,250, 3,125, etc. cells/well) into an empty 24-well plate, aliquoting CSC into 500 μL of Cell Dissociation Solution.
   (b) Add 12 μL of Calcein-AM Solution to 12 mL of Cell Dissociation Solution. Mix well.
   (c) Add 500 μL of Cell Dissociation Solution/Calcein-AM to each set of wells containing decreasing numbers of cells, and incubate for 1 h; omit cells from at least three wells to calculate background.
   (d) Read at 485 nm excitation, 520 nm emission to obtain RFU values.
   (e) Average your values for each cell concentration; then subtract the background from each value.
   (f) Plot standard curve RFU values vs. number of cells.
   (g) Insert a trend line (best fit) and use the equation $y = ax + b$ for each cell line to calculate the number of cells that invaded.
5. Perform the migration experiment as follows: for each replicate, add 100 μL of cell suspension to the upper chamber and 500 μL of BM added with 10% FBS to the lower chamber. To compensate for background, instead of adding cell suspension, add 100 μL of BM in at least three replicates.
6. Incubate at 37°C, 5% CO_2 for 10 h.
7. After incubation, aspirate top chamber, remove unmigrated cells by gently rubbing the filter membrane with a cotton swab, and wash each well with 100 μL warm HBSS.
8. Aspirate each bottom chamber, and wash them twice with 500 μL warm HBSS.
9. Add 12 μL of Calcein AM solution to 12 mL of Cell Dissociation Solution.
10. Add 500 μL of Cell Dissociation Solution/Calcein AM to the bottom chamber of each well, reassemble the chambers, and incubate them at 37°C, 5% CO_2 for 30 min.
11. Gently but firmly tap the plate on the side, and incubate it at 37°C, 5% CO_2 for an additional 30 min.
12. Remove inserts and read plate at 485 nm excitation, 520 nm emission using same parameters (time and gain) as standard curve, or controls.

13. Compare experimental data to controls, and convert RFU into cell number to determine the number of cells that have migrated.

4 Notes

1. In order to scale down the assays and minimize cell requirement, glass coverslips in 24-well plates can be substituted with 96-well imaging plates (e.g., BD high-content imaging plates). The thin bottom of these plates provides high light transmittance and excellent optics for imaging. This may be coupled with semi-automated imaging systems for high-content screening purposes (e.g., BD Pathway Bioimager 435 or 855, Opera High Content Screening System-Perkin Elmer).
2. For best results use paraformaldehyde (PFA) within few days from preparation. It may be kept in the dark at +4°C for 4–5 days. For longer storage, prepare aliquots freeze them and keep them at −20°C. Do not use the pH meter to check its pH since PFA may ruin the electrode.
3. Most antibodies tested for immunohistochemistry or immunofluorescence will work on PFA fixed cells, however aldehyde fixatives may reduce antigen reactivity.
4. The main advantage of employing secondary antibodies all of which are raised in the same animal species is given by the reduced cross-reactivity between secondary antibodies. Furthermore, a Fab fragment of the primary antibody may be used instead of the whole antibody to further reduce recognition of secondary antibodies via their Fc region, and to increase their penetration through a smaller size.
5. Fixation may be reduced to 10 min for antigens expressed on the cell surface, since a longer one could reduce their antigen reactivity.
6. Antifading agents are useful in reducing the photo-bleaching of fluorescently labeled cells, probably by scavenging free radicals. However, they could reduce fluorescence intensity. Several ready to use agents are available, therefore it is recommended to choose the most appropriate one for the application of choice.
7. Denaturation and hybridization steps may be carried out in an incubator. Best results are obtained by preheating the incubator to the appropriate temperature and placing the slides on a flat metal plate located in the incubator. As an alternative, automated devices, such as the Thermobrite (Abbot, USA) have several advantages, such as rapid temperature ramp-up and accuracy of ±1°C, temperature uniformity across all slide positions, and humidity control.
8. With respect to the protocol provided by the company, we observed that an overnight hybridization results in a stronger signal, which facilitates the analysis of dysfunctional telomeres.

Aknowledgement

Italian Ministry of Health. GR-2007-683407.

References

1. Beltrami AP, Cesselli D, Beltrami CA (2012) Stem cell senescence and regenerative paradigms. Clin Pharmacol Ther 91:21–29
2. Beltrami AP, Cesselli D, Beltrami CA (2011) At the stem of youth and health. Pharmacol Ther 129:3–20
3. Rodier F, Campisi J (2011) Four faces of cellular senescence. J Cell Biol 192:547–556
4. Mandal PK, Blanpain C, Rossi DJ (2011) DNA damage response in adult stem cells: pathways and consequences. Nat Rev Mol Cell Biol 12:198–202
5. Rossi DJ, Jamieson CH, Weissman IL (2008) Stems cells and the pathways to aging and cancer. Cell 132:681–696
6. Sharpless NE, DePinho RA (2007) How stem cells age and why this makes us grow old. Nat Rev Mol Cell Biol 8:703–713
7. Campisi J (2005) Senescent cells, tumor suppression, and organismal aging: good citizens, bad neighbors. Cell 120:513–522
8. Minamino T, Komuro I (2008) Vascular aging: insights from studies on cellular senescence, stem cell aging, and progeroid syndromes. Nat Clin Pract Cardiovasc Med 5:637–648
9. Torella D, Rota M, Nurzynska D, Musso E, Monsen A, Shiraishi I, Zias E, Walsh K, Rosenzweig A, Sussman MA, Urbanek K, Nadal-Ginard B, Kajstura J, Anversa P, Leri A (2004) Cardiac stem cell and myocyte aging, heart failure, and insulin like growth factor-1 overexpression. Circ Res 94:514–524
10. Urbanek K, Torella D, Sheikh F, De Angelis A, Nurzynska D, Silvestri F, Beltrami CA, Bussani R, Beltrami AP, Quaini F, Bolli R, Leri A, Kajstura J, Anversa P (2005) Myocardial regeneration by activation of multipotent cardiac stem cells in ischemic heart failure. Proc Natl Acad Sci USA 102:8692–8697
11. Campisi J, d'Adda di Fagagna F (2007) Cellular senescence: when bad things happen to good cells. Nat Rev Mol Cell Biol 8:729–740
12. Jeyapalan JC, Sedivy JM (2008) Cellular senescence and organismal aging. Mech Ageing Dev 129:467–474
13. Blasco MA (2007) Telomere length, stem cells and aging. Nat Chem Biol 3:640–649
14. Deng Y, Chan SS, Chang S (2008) Telomere dysfunction and tumour suppression: the senescence connection. Nat Rev Cancer 8:450–458
15. Shawi M, Autexier C (2008) Telomerase, senescence and ageing. Mech Ageing Dev 129:3–10
16. Chimenti C, Kajstura J, Torella D, Urbanek K, Heleniak H, Colussi C, Di Meglio F, Nadal-Ginard B, Frustaci A, Leri A, Maseri A, Anversa P (2003) Senescence and death of primitive cells and myocytes lead to premature cardiac aging and heart failure. Circ Res 93:604–613
17. Campisi J (2011) Cellular senescence: putting the paradoxes in perspective. Curr Opin Genet Dev 21:107–112
18. Cesselli D, Beltrami AP, D'Aurizio F, Marcon P, Bergamin N, Toffoletto B, Pandolfi M, Puppato E, Marino L, Signore S, Livi U, Verardo R, Piazza S, Marchionni L, Fiorini C, Schneider C, Hosoda T, Rota M, Kajstura J, Anversa P, Beltrami CA, Leri A (2011) Effects of age and heart failure on human cardiac stem cell function. Am J Pathol 179:349–366
19. Lawless C, Wang C, Jurk D, Merz A, Zglinicki T, Passos JF (2010) Quantitative assessment of markers for cell senescence. Exp Gerontol 45:772–778
20. Lawless C, Jurk D, Gillespie CS, Shanley D, Saretzki G, von Zglinicki T, Passos JF (2012) A stochastic step model of replicative senescence explains ROS production rate in ageing cell populations. PLoS One 7:e32117
21. d'Adda di Fagagna F (2008) Living on a break: cellular senescence as a DNA-damage response. Nat Rev Cancer 8:512–522
22. Fumagalli M, Rossiello F, Clerici M, Barozzi S, Cittaro D, Kaplunov JM, Bucci G, Dobreva M, Matti V, Beausejour CM, Herbig U, Longhese MP, d'Adda di Fagagna F (2012) Telomeric DNA damage is irreparable and causes persistent DNA-damage-response activation. Nat Cell Biol 14:355–365
23. Herbig U, Ferreira M, Condel L, Carey D, Sedivy JM (2006) Cellular senescence in aging primates. Science 311:1257
24. Baerlocher GM, Vulto I, de Jong G, Lansdorp PM (2006) Flow cytometry and FISH to measure the average length of telomeres (flow FISH). Nat Protoc 1:2365–2376
25. Vindelov LL, Christensen IJ (1990) A review of techniques and results obtained in one laboratory by an integrated system of methods designed for routine clinical flow cytometric DNA analysis. Cytometry 11:753–770

Chapter 8

Isolation of Mesenchymal Stem Cells from Human Bone and Long-Term Cultivation Under Physiologic Oxygen Conditions

Sebastian Klepsch, Angelika Jamnig, Daniela Trimmel, Magdalena Schimke, Werner Kapferer, Regina Brunauer, Sarvpreet Singh, Stephan Reitinger, and Günter Lepperdinger

Abstract

Bone-derived stroma cells contain a rare subpopulation, which exhibits enhanced stemness characteristics. Therefore, this particular cell type is often attributed the mesenchymal stem cell (MSC). Due to their high proliferation potential, multipotential differentiation capacity, and immunosuppressive properties, MSCs are now widely appreciated for cell therapeutic applications in a multitude of clinical aspects. In line with this, maintenance of MSC stemness during isolation and culture expansion is considered pivot. Here, we provide step-by-step protocols which allow selection for, and in vitro propagation of high quality MSC from human bone.

Keywords Bone marrow, Mesenchymal stem cells, Isolation, Growth curve

1 Introduction

Mesenchymal stem cells (MSC), a subpopulation of multipotent mesenchymal stromal cells (1), are tissue-specific adult stem cells involved in regeneration and homeostasis (2, 3), and found in all stromal tissues throughout the body (4, 5). There, they are thought to reside in niches, which allow regulated dormancy, as well as controlled production of progeny to foster tissue regeneration. In the bone marrow, MSC appear to share a common niche with hematopoietic stem cells (6).

Many research laboratories use bone marrow aspirate as a rich source to attain this particular cell type for subsequent cell culture experiments. Presently, MSC are isolated and cultured following different methods and protocols (7–10). Common characteristics of MSC could only be insufficiently specified, first and foremost because the cell isolates are heterogeneous and comprise more than

Kursad Turksen (ed.), *Stem Cells and Aging: Methods and Protocols*, Methods in Molecular Biology, vol. 976, DOI 10.1007/978-1-62703-317-6_8,

one mesenchymal cell type. It is therefore difficult to resolve how results regarding ex vivo propagated cells relate to each other.

A large body of information is available on many aspects of cultivated MSC properties first described in (11): they firmly adhere to cell culture plastic, which is a distinctive functional criterion for selection during isolation; they furthermore exhibit clonogenic growth, which is another distinctive criterion as during expansion low density seeding selects for rapidly proliferating cells; moreover, unique and highly specialized cells within the heterogeneous population of mesenchymal stromal cells show, when cultured under specific conditions, manifold differentiation capabilities in vitro (12).

In this context, questions of how to select high quality MSC with respect to basic stem cell functions, or what are appropriate surface markers and in vitro conditions for propagation of MSC for later clinical application, and lastly, which systemic factors actually impinge on their respective differentiation capacity need to be addressed with the prime aim to eventually establish standardized protocols in order to achieve comparability of results between research laboratories (12, 13).

To this end, no single protocol has proven superior over others. Thus, consolidation appears difficult. The isolation and propagation protocol introduced here addresses two challenges. First challenge: often bone marrow is aspirated and passed on to further purification disregarding the fact that the suspension contains clumpy material and minuscule bony fragments. Treatment with collagenase allows gentle disintegration of extracellular matrix and thus opening of the MSC niche. In this way significantly more MSC-like cells can be obtained (8). Second challenge: similar to most somatic cell types, MSC eventually exhibit irreversible growth arrest also called senescence, a process paralleled by progressive telomere shortening (14); intermediate low-oxygen tension or physiologically normoxic conditions during in vitro cultivation allows prolonged MSC expansion while at the same time assuring multipotential differentiation capacity (8).

2 Materials

2.1 Laboratory Hardware for Cell Culture

1. Sterile laminar flow hood.
2. Regular cell culture incubator (20% pO_2, 5% pCO_2).
3. Low oxygen (3% pO_2) incubator: Thermo Scientific Forma® Model 3131 Series II Water Jacketed CO_2 Incubator Class 100 (see Note 1).
4. Centrifuge: equipped with swing-out rotor capable of accommodating 15-ml tubes.
5. Neubauer improved hemocytometer (#442/72; Assistant).

6. Inverted microscope equipped with 5×, 10×, and 20× objectives.
7. Freezing container (#5100; NALGENE).
8. Sterile pair of scissors and tweezers (see Note 2).

2.2 Disposables, Solutions, and Buffers

1. Cell culture plastic: all sterile: (see Note 3).
   (a) 50-ml centrifuge tubes, V-shaped bottom.
   (b) 15-ml centrifugation tubes, V-shaped bottom.
   (c) 0.5-ml reaction tubes without cap.
   (d) 2-ml cryo/freezing tubes.
   (e) 1-ml disposable plastic Gilson-type pipette tips.
   (f) Cell culture dishes 35×10 mm.
   (g) Cell culture dishes 100×20 mm.
   (h) Cell strainers (nylon, 100 μm mesh size).
   (i) 24-well cell culture plates.
   (j) 6-well cell culture plates.
2. Growth medium: minimal essential medium (MEM)+ Glutamax™-I (Gibco #41090; Invitrogen), 20% heat-inactivated fetal calf serum (FCS), 100 units/ml penicillin, and 100 μg/ml streptomycin (see Note 4).
3. Purified collagenase (#LS005273; Worthington) (see Note 5).
4. Density gradient: Ficoll-Paque™ PLUS, 1.077 g/ml (#17-1440; GE Healthcare).
5. Cell staining: 0.25% Trypan Blue in phosphate-buffered saline (PBS).
6. Cell splitting: 0.05% Trypsin/1 mM Ethylenediaminetetraacetic acid (EDTA).
7. Dulbecco's-modified phosphate-buffered saline, DPBS.
8. Freeze medium: MEM+Glutamax™-I, 30% heat-inactivated fetal calf serum (FCS), 5% dimethyl sulfoxide (DMSO), 100 units/ml penicillin, and 100 μg/ml streptomycin.

3 Methods

3.1 Isolation of Bone Stromal Cells

All procedures, in particular when concerning specific scientific research questions or clinical applications employing human MSC, have to be approved by an Institutional Review Board, and donors have to provide their written informed consent prior to surgery.

The here described method was developed using a small biopsy (<1 cm^3) of cancellous bone (i.e., *substantia spongiosa ossium*) from the iliac crest obtained from systemically healthy individuals who underwent reconstructive bone surgery due to craniomaxillofacial indications. Post operation, the remaining bone pieces would have been discarded otherwise. After surgery, all steps were performed in a cell culture laboratory equipped with a sterile laminar flow hood.

1. Prior to surgery, fill a sterile 50-ml centrifuge tube with 25 ml growth medium. The container can be also stored for prolonged periods at 4°C, but should be warmed to ambient temperature before biopsy collection and transport.
2. Provide immediate transportation to the cell culture laboratory; take care to keep biopsies at room temperature.
3. For subsequent manipulation of the spongious sample, transfer the container into the sterile hood. Use a sterile pair of tweezers to transfer bone fragments into a 35 × 10 mm sterile culture dish (see Note 6).
4. Cover bone with growth medium (Fig. 1d(*)), and use a sterile pair of scissors to downsize the bone piece to 20–100 mm^3 fragments; the pieces should later (see steps 6 and 7) smoothly slip into a 1-ml Gilson-type plastic pipette tip (see Note 7).
5. Transfer bone biopsies into a sterile 35 × 10 mm culture dish and add 4 ml growth medium containing 20 units/ml purified collagenase pre-warmed to 37°C; incubate at 37°C, 20% O_2, 5% CO_2 for 2 h.
6. Assemble cell isolation centrifugation appliance(s) as depicted in Fig. 1a(i), comprising a 1-ml plastic pipette with a truncated tip, a 0.5-ml disposable reaction tube, and a 15-ml centrifugation tube (10) (see Note 8).
7. Place bone fragments into the pipette tip (Fig. 1b, also read Note 9) and separate loosened bulky material from spongious bone by centrifugation at 400 × *g* for 1 min at room temperature.
8. Discard bone fragments together with the 1-ml pipette tip (see Note 10); follow institutional instruction for disposal of biohazard waste.
9. Resuspend cell pellet in 500 μl growth medium using a 1-ml syringe mounted with a 0.80 × 40 mm/21 G needle. Transfer cell suspensions to a 15-ml tube; make pools of up to 5 ml. Add growth medium to a final volume of 5 ml.
10. Remove miniscule bone fragments floating in the cell suspension by filtering through a cell strainer into a 35 × 10 mm culture dish (see Note 11).
11. Aspire the cleared cell suspension into a plastic pipette; carefully layer 5 ml above 5-ml Ficoll-Paque™ in a 15-ml centrifugation tube (Fig. 1c, d(iii), see Note 12) and spin at 2,500 × *g* for 30 min at room temperature in a centrifuge; make sure that brakes are deactivated.

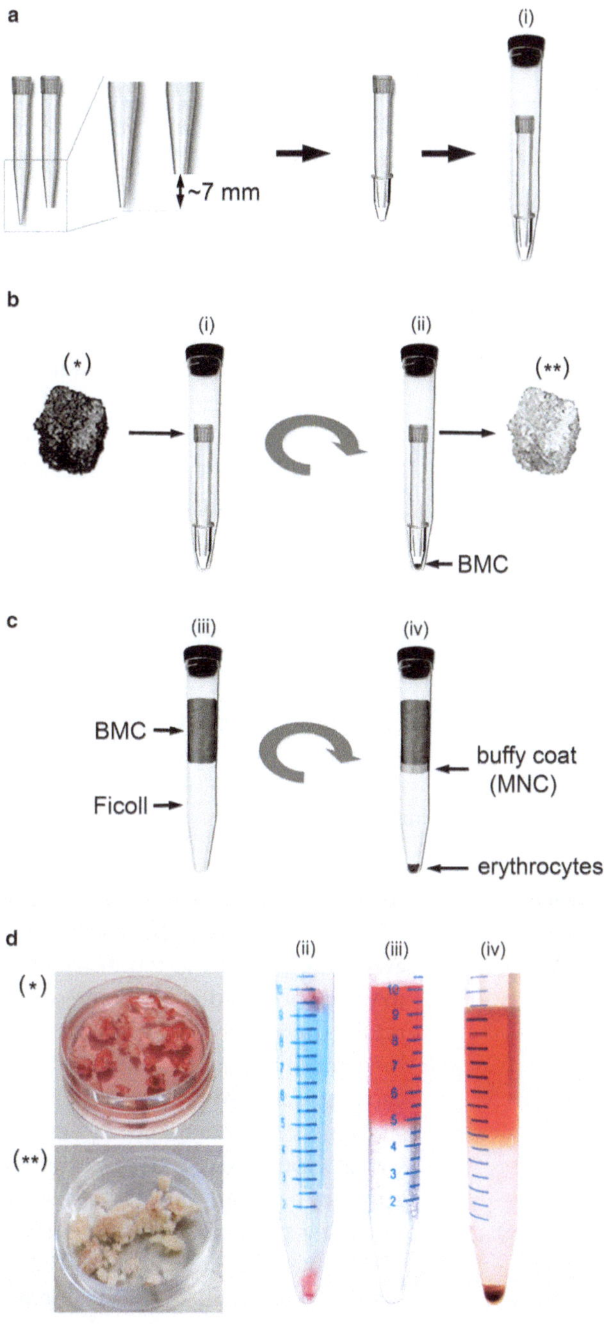

Fig. 1 MSC isolation: (**a**) Assembled cell isolation centrifugation appliance (i) consisting of a truncated pipette tip, a 0.5-ml vial, and a 15-ml tube. (**b**) Collagenase-digested bone specimen (*asterisk*) is placed into pipette tip of the centrifugation appliance (i). During centrifugation (ii) bone marrow cells (BMC) are gathered at the bottom of the 0.5-ml vial. The cell-depleted bone fragment (*double asterisk*) is discarded. (**c**) Ficoll-Paque™ is overlaid with BMC diluted in growth medium (iii). Density gradient centrifugation separates dense erythrocytes from mononuclear cells (MNC), the latter forming a turbid interphase layer also called buffy coat (iv). (**d**) Representative photographs showing individual isolation steps, appliances, and results

12. Carefully isolate the interphase, also called "buffy coat," comprising most mononuclear cells (MNC), thus also including MSC (Fig. 1c, d(iv)) (see Note 13).
13. Add 10 ml growth medium to the isolated interphase in order to dilute Ficoll-Paque™. Spin cells down at 1,500×*g* for 15 min at room temperature. Carefully aspirate the supernatant; take care to leave the cell pellet undisturbed. Resuspend cell pellet in 1 ml growth medium.
14. Assessment of cell number: Add 80 ml of Trypan Blue solution to 20 ml of (diluted) cell suspension in a 96 well plate and mix well. Transfer 10 ml of stained cells into a Neubauer improved hemocytometer. Count cells at 100-fold magnification (10× objective). Dead cells stain deep blue, while viable cells remain bright. Count the number of living cells in at least four quadrants and calculate the mean thereof. Determine cell concentration (cell/ml) by multiplying average count of viable cells by 5 (= dilution factor in Trypan Blue) and 10,000 (= volume in the counting chamber).
15. Plate 200,000 cells/cm^2 and incubate at 37°C, 3% O_2, 5% CO_2 overnight.
16. On the next day, carefully remove non-adherent cells by gently rinsing with growth medium. Thereafter wash adherent cells twice with 15 ml pre-warmed DPBS (see Note 14).
17. Expand cells; for detailed instructions see next section.

3.2 Cultivation of MSC

Mesenchymal stroma cells contain a rare population bearing stem cell-like properties. Selection for this particular cell type can be achieved through low density seeding, thus provoking colony formation and rapid clonogenic proliferation. In order to enforce and retain these stemness characteristics, cells are always (re)seeded at low density (50 cells/cm^2). To determine long-term proliferative capacity cells are propagated in a 100×20 mm dish. The low density cultures are expanded, yet utmost care is taken to keep individual clones nicely demarcated by not letting them grow into a confluent monolayer. During early passages, low density cultures normally reach a subconfluent state for further passaging within 10–14 days, at later passages this period increases. Expansion time is by and large also donor dependent. Often during MSC cultivation transition from spindle-shaped, fibroblast-like cell morphology, typical for early passages, to large and flattened shape characteristic for replicatively senescent cells in late passages is observed frequently (Fig. 2).

For subcultivation, the following protocol is processed (the stated volumes refer to a 100×20 mm culture dish):

1. Before starting with cell manipulation, pre-warm growth medium, DPBS and Trypsin/EDTA in a water bath to 37°C;

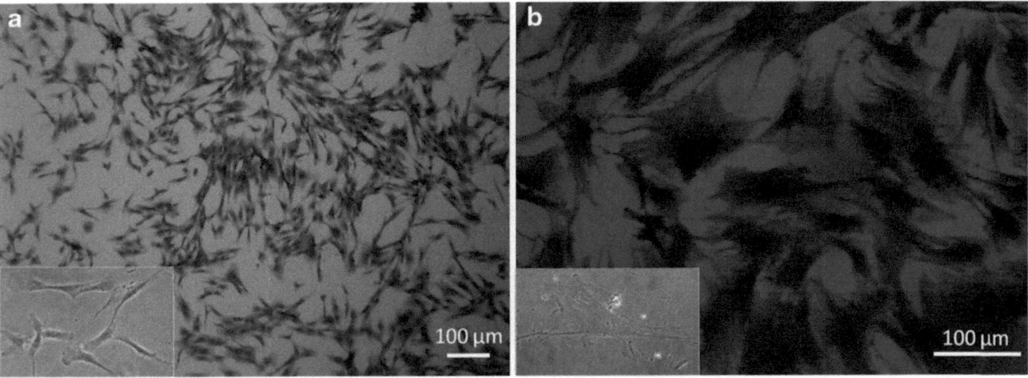

Fig. 2 Phenotypic appearance of cultured MSC (**a**) slender, spindle-shaped phenotype of plastic adherent MSC at an early passage (P2); cells were stained with crystal violet to clearly visualize the cellular morphology, image detail shows a close-up of a few living cells at the same passage by means of phase contrast microscopy. (**b**) typical flattened morphology of replicatively aged MSC at late passage (P9) exhibiting increased cell volume

take care that the temperature of all solutions are equally well adjusted to 37°C.

2. Aspirate culture medium (see Note 15).
3. Add 4 ml DPBS and incubate at 37°C for 5 min to remove residual medium and discard supernatant. Repeat this step.
4. Add 2 ml Trypsin/EDTA and incubate at 37°C for 5 min (see Note 16).
5. Detach cells from the plastic surface by forcefully knocking the dish sideways against the bench; examine rate of detachment under the microscope.
6. Add 2 ml growth medium to the cell suspension; support further detachment by carefully pipetting cells up and down while holding the dish lopsided to concentrate most cells at one side of the dish (see Note 17).
7. Transfer cell suspension to a 15-ml centrifugation tube; rinse the culture dish with another 5 ml of growth medium to collect all remaining cells; spin at 400 × g for 1 min at room temperature.
8. Aspirate and discard supernatant; take care not to disturb the cell pellet;
9. Resuspend cells in 1 ml growth medium and determine the cell number as described in step 14, Subheading 3.1.
10. To assess cumulative population doublings in long-term proliferation experiments, use the following formula

$$\frac{\ln(\text{terminal cell number} / \text{number of seeded cells})}{\ln(2)}$$

Progressive change in proliferative potential is best displayed by presenting accumulation of population doublings over time.

3.3 Long-Term Storage

Prior to storage, MSC need to be detached from culture dishes following steps 1–9 of the previous section.

3.3.1 Freezing

1. Resuspend cell pellet after centrifugation at 400 × *g* for 1 min at RT in growth medium (volume) and determine the cell number as outlined in step 14, Subheading 3.1.
2. Spin the cells down and dissolve in 4°C cold freeze medium to 1×10^6 cells/ml. Ensure that cells are kept at 4°C from this point on (see Note 18).
3. Fill not more than 1 ml of cell suspension into a single vial, which needs to be specially designed for cryopreservation, and place the cryovials into a freezing container (conditioned at 4°C).
4. Place container together with the vials for 24 h into a −80°C freezer; thereafter transfer the vials into liquid nitrogen for long-term storage.

3.3.2 Thawing and Recultivation

1. Immediately after retrieval from liquid nitrogen, thaw the frozen cell suspension in a 37°C water bath; shortly before thawing is completed dilute suspension in 10 ml 4°C cold growth medium.
2. Centrifuge at 400 × *g* for 1 min at room temperature.
3. Discard the supernatant and resuspend the pellet in 1 ml 37°C warm growth medium.
4. For assessment of cell numbers, follow step 14, Subheading 3.1.
5. Seed 2,000 cells/cm^2 (see Note 19).

4 Notes

1. Cultivation under normoxic conditions requires a specially devised incubator. Oxygen tension can be adjusted with the aid of a built-in oxygen sensor which controls nitrogen inlet thereby displacing atmospheric oxygen by nitrogen. Nitrogen used herein had a purity of >99.999% containing <2 ppm O_2, <3 ppm H_2O or <0.1 hydrocarbon (methane). Concomitantly, CO_2 content is controlled via a CO_2 sensor and injection of medical grade CO_2.
2. Instruments have to be carefully cleaned and steam-sterilized prior to use.
3. Manufacturer-dependent variations in the surface processing and finishing of cell culture plastic may have an influence on basic cell properties. It is therefore highly recommended to test plastic ware for best results regarding MSC adherence, colony formation, and differentiation (not specified in this

protocol). Protocols described herein were established using products purchased from Sarstedt (Sarstedt GmbH, Wiener Neudorf, Austria) and BD (Becton Dickins Austria GmbH, Schwechat, Austria).

4. The quality of fetal calf serum has to be tested and those batches selected, which warrant undisturbed cell growth; one may also consider testing the supplements whether differentiation capacity or immune-modulatory potential of MSC is compromised (not specified in this protocol).
5. Purified collagenase is dissolved in growth medium and stored as 100 μl aliquots of 1,000 units/ml. Apply 20 units/ml final concentration of purified collagenase. The use of purified collagenase is mandatory when using MSC for subsequent coculture with immune cells as trace contaminations present in most of the conventional enzyme formulations isolated from bacterial cultures will activate the cells and impact on biological processes thus corrupting specific experimental approaches later on.
6. The size of the tweezers should allow easy handling, grabbing, and recapturing of material within the 50-ml transportation container. Take utmost care not to contaminate the instruments before all bony pieces are transferred to the dish.
7. Use as much growth media as necessary to cover the bone fragments almost completely. Use tweezers to pin down a bone biopsy during fragmentation with scissors. Take care during splitting as during manipulation fragments may dart off and pop out of the dish.
8. Widen the opening of a 1-ml Gilson-type plastic pipette by truncating its tip by ~7 mm with a pair of scissors prior to autoclaving. Assemble the separation appliance as depicted in Fig. 1a. Take care that the pipette tip sits firmly in the 0.5 ml collection vial; the tip should be well supported by the tube wall at a distance of ~1 mm above the base of the vial so that the opening is not clogged during centrifugation.
9. Place the fragments into the pipette so that the pieces are well trapped, and thus are forced down towards the tapering opening during centrifugation (Fig. 1(iii) d(ii)). It is recommended not to place too many fragments into a single tip but rather use several appliances instead.
10. After centrifugation, search for a reddish pellet at the bottom of the 0.5-ml vial. Remove the bone fragment containing pipette and inspect the individual fragments. A palish, bright appearance indicates sufficient separation of bulky cellular material from the spongious bone pieces (Fig. 1d(**)).
11. After filtration through the cell strainer, often a large hanging drop forms underneath, which needs to be carefully collected together with the flow-through.

12. Prepare the Ficoll gradient in a 15-ml centrifugation tube (V-shaped bottom). First aspirate 5 ml Ficoll-Paque™ from the storage bottle with the aid of a syringe. Carefully overlay the dense Ficoll phase with 5 ml cell suspension using a 5-ml pipette (Fig. 1d(iii)); mixing of the two layer can be avoided by tilting the 15-ml tube to gain better access to the enlarged interphase surface and also by slowly dribbling the cell suspension on top of the Ficoll layer.
13. For collection of buffy coat (Fig. 1c, d(iv)), use a 1-ml pipette and discreetly penetrate the upper reddish media layer. Take care to collect all of the turbid interphase without any collateral Ficoll underneath. Aspirate up to 1 ml of buffy coat and pool in a sterile 15-ml tube.
14. MSC will firmly attach to the plastic surface, whereas most other MNC keep floating; in this state, these can be easily removed. Usually, mesenchymal stromal cells adhere to the plastic surface 24 h after seeding.
15. For aspiration of the medium, use a glass pipette connected via flexible silicone tubing to a container which is evacuated with the aid of a vacuum pump. Do not touch the cellular monolayer; slightly tilt the dish so that media collects at one point and thus can be more easily aspirated without disturbing the cells.
16. Check detachment of cells under a microscope. Only when a majority of the cells shows a roundish phenotype, or is floating in suspension, stripping of the cells from the plastic surface was sufficient. In case cells still adhere and display fibroblast-like appearance, continue incubation at 37°C and recheck under the microscope at least every other minute, yet do not exceed 10 min of trypsin treatment.
17. Serum contains trypsin inhibitors and thus growth medium will render trypsin inactive almost immediately. Take care to collect all cells by flushing the solution over the plastic surface and thus clearing the plastic dish from most of cells.
18. DMSO is used as a cryoprotectant, which readily crosses the cell wall and binds to water molecules intracellularly. Complexed in this way the efflux of water from the cytoplasm is blocked during freezing, thereby preventing cellular dehydration or shrinkage and thus maintaining stable intracellular salt concentrations and pH levels. By slowing the rate of freezing, DMSO eventually prevents harmful ice crystals formation within the cell. It has been reported for various cell lines that DMSO concentrations higher than 10% may exert toxic side effects, in particular when treated at 37°C. It is therefore imperative to keep cells chilled at 4°C, and notably after thawing, to thoroughly rinse cells with growth medium.

19. In case of cell seeding after long-term storage, culture is restarted at an initial density of 2,000 cell/cm^2 instead of 50 cell/cm^2 as performed under routine conditions.

Acknowledgment

SR is supported by a Marie Curie International Reintegration Grant. GL is supported by research funds granted by the Austrian Research Agency FFG—Laura Bassi Centre of Expertise DIALIFE, the Translational Research Programme of the Tyrolean Future Fund, and the EC's 7th framework program VascuBone FP7-HEALTH-2009-single-stage-242175 coordinated by Heike Walles, University of Würzburg.

References

1. Horwitz EM, Le Blanc K, Dominici M et al (2005) Clarification of the nomenclature for MSC: the International Society for Cellular Therapy position statement. Cytotherapy 7:393–395
2. Valtieri M, Sorrentino A (2008) The mesenchymal stromal cell contribution to homeostasis. J Cell Physiol 217:296–300
3. Phinney DG, Prockop DJ (2007) Concise review: mesenchymal stem/multipotent stromal cells: the state of transdifferentiation and modes of tissue repair—current views. Stem Cells 25:2896–2902
4. Gamblin SJ, Davies GJ, Grimes JM et al (1991) Activity and specificity of human aldolases. J Mol Biol 219:573–576
5. Crisan M, Yap S, Casteilla L et al (2008) A perivascular origin for mesenchymal stem cells in multiple human organs. Cell Stem Cell 3:301–313
6. Mendez-Ferrer S, Michurina TV, Ferraro F et al (2010) Mesenchymal and haematopoietic stem cells form a unique bone marrow niche. Nature 466:829–834
7. D'Ippolito G, Diabira S, Howard GA et al (2004) Marrow-isolated adult multilineage inducible (MIAMI) cells, a unique population of postnatal young and old human cells with extensive expansion and differentiation potential. J Cell Sci 117:2971–2981
8. Fehrer C, Brunauer R, Laschober G et al (2007) Reduced oxygen tension attenuates differentiation capacity of human mesenchymal stem cells and prolongs their lifespan. Aging Cell 6:745–757
9. Gronthos S, Graves SE, Ohta S et al (1994) The STRO-1+ fraction of adult human bone marrow contains the osteogenic precursors. Blood 84:4164–4173
10. Peister A, Mellad JA, Larson BL et al (2004) Adult stem cells from bone marrow (MSCs) isolated from different strains of inbred mice vary in surface epitopes, rates of proliferation, and differentiation potential. Blood 103: 1662–1668
11. Friedenstein AJ, Petrakova KV, Kurolesova AI et al (1968) Heterotopic of bone marrow. Analysis of precursor cells for osteogenic and hematopoietic tissues. Transplantation 6: 230–247
12. Dominici M, Le Blanc K, Mueller I et al (2006) Minimal criteria for defining multipotent mesenchymal stromal cells. The International Society for Cellular Therapy position statement. Cytotherapy 8:315–317
13. Bianco P, Robey PG, Simmons PJ (2008) Mesenchymal stem cells: revisiting history, concepts, and assays. Cell Stem Cell 2:313–319
14. Simonsen JL, Rosada C, Serakinci N et al (2002) Telomerase expression extends the proliferative life-span and maintains the osteogenic potential of human bone marrow stromal cells. Nat Biotechnol 20:592–596

Chapter 9

Methods for Assessing Effects of Wnt/β-Catenin Signaling in Senescence of Mesenchymal Stem Cells

Hai-jie Wang and Yu-zhen Tan

Abstract

Mesenchymal stem cells (MSCs) represent a main population of stem cells and can differentiate into multiple cell lineages. Recently, MSC transplantation has been applied to repair the malfunctioned tissues. However, increasing evidences show that some MSCs expanded in vitro and in the aged individuals become senescent. Capacity of senescent MSCs in repairing the tissues may decrease significantly. Interestingly, preventing MSC senescence is a powerful potential strategy to delay aging of individuals and promote application of cell therapy for treating aging-related diseases. Therefore, it is necessary to explore mechanisms of MSC senescence in detail. Methods to assess MSC senescence in vitro include induction of senescence, detection of senescent changes and investigation of the molecules involved in senescence. Here we describe the methods to detect MSC senescence induced with old serum and investigate effects of Wnt/β-catenin signaling on MSC senescence.

Keywords Cellular senescence, Mesenchymal stem cells, SA-β-gal, Wnt/β-catenin signaling, DNA damage response, p53, p21, p16^{INK4a}

1 Introduction

Cells continually experience stress and damage from exogenous and endogenous sources, and their responses range from complete recovery to cell death. The damage may be completely repaired, restoring the cell and tissue to its pre-damaged state. Excessive or irreparable damage, however, can cause senescence, apoptosis, or an oncogenic mutation (1). Cellular senescence is a form of irreversible cell cycle arrest and can be induced by heritable telomere depletion, exogenous DNA damage, aging niche, or oncogene overexpression (2–5). Decline of regenerative capacity of stem cells appears to contribute to human age-associated conditions such as frailty, atherosclerosis, and type 2 diabetes (2, 6).

Kursad Turksen (ed.), *Stem Cells and Aging: Methods and Protocols*, Methods in Molecular Biology, vol. 976,
DOI 10.1007/978-1-62703-317-6_9, © Springer Science+Business Media, LLC 2013

Adult stem cells are responsible for maintaining tissue homeostasis throughout an organism's lifetime, thus aging-related phenotypes might be, at least in part, due to a decline in the number or function of tissue stem cells (3). Mesenchymal stem cells (MSCs) represent a main population of stem cells besides embryonic stem cells and haemopoietic stem cells. MSCs are present in a variety of tissues during human development, and in adults they are prevalent in bone marrow and also reside around blood vessels, in skin, fat, muscle, and other locations (7). Bone marrow is viewed as the richest and most reliable reservoir for MSCs (8). Under induction with cytokines, MSCs differentiate toward osteoblasts, chondrocytes, myoblasts, stromal cells of bone marrow, fibroblasts, adipocytes, dermal cells, etc. and fabricate a spectrum of specialized mesenchymal tissues. MSCs in human bone marrow decrease with age (9). There is an age-dependent decrease in proliferation and osteoblast differentiation, and an increase in senescence-associated β-galactosidase(SA-β-gal)-positive cells (10). Compared with young cells, the aged MSCs display a large, flattened, and no spindle-formed morphology. The senescent cells exhibit more podia, spread further, and contain more actin stress fibers (11). Lysosomes and autolysosomes increase in the senescent cells (12).

In recent years, more and more attention focuses on exploring molecular mechanisms of MSC senescence. There are two major pathways for induction of cellular senescence. One is telomere-dependent pathway, which is mediated by p53/p21 signaling pathway and DNA damage response. Another is telomere-independent pathway activated by oxidative stress via regulation of Erk-$p38^{MAPK}$ signaling pathway (13). In addition, accumulation of cell cycle inhibitor proteins like $p16^{INK4a}$ and heat shock proteins contribute to cellular senescence (9, 14). Reactive oxygen species (ROS) may cause DNA damage. Outcomes of DNA damage response include transient cell cycle arrest coupled with DNA repair, proliferation, senescence, or apoptosis (15). DNA methyltransferases play a critical role in regulating MSC senescence through controlling not only the DNA methylation status but also active/inactive histone marks at genomic regions of PcG-targeting miRNAs and $p16^{INK4A}$ and $p21^{CIP1/WAF1}$ promoter regions (16). Moreover, ROS induces premature senescence of human MSCs by regulating some proteins and metabolites. Significant and possible senescence-related proteins or metabolites are Annexin A2, prosomal protein P30-33K and glycine, proline, choline, and leucine (17).

The most interesting evidence shows that aging systemic milieu negatively regulates cellular senescence. The age related decline of stem cell activity can be modulated by systemic factors (18, 19). Recently, we have examined the effects of old rat serum (ORS) on MSC aging, and explored mechanisms of Wnt/β-catenin signaling on ORS-induced MSC aging. ORS promotes MSC senescence and reduces proliferation and survival of the cells. After treatment with ORS, Wnt/β-catenin signaling is activated. Expression of γ-H2A.X (marker of DNA damage response), $p16^{INK4a}$, p53, and p21 is increased in ORS-induced senescent MSCs. These results suggest that Wnt/β-catenin signaling plays a critical role in MSC aging induced by factors in ORS. DNA damage response and p53/p21 pathway are main mediators of MSC aging mediated with Wnt/β-catenin signaling (20). Key serum factors inducing MSC aging may be potential targets for anti-aging.

Cultures of MSCs have been successfully established from bone marrow, umbilical cord blood, trabecular bone, periosteum, synovium, placenta, pancreas, adipose tissue, skin, lung, and thymus. Application of bone marrow-derived MSCs has an excellent perspective in cell therapy, tissue engineering and regenerative medicine than embryonic stem cells that restricted in clinical application owing to their immunogenicity and ethical issues. Transplantation of MSCs has a clinical potential for tissue repair in some diseases, such as myocardial infarction (21–23). However, morrow-derived MSCs are rare so that the cells have to be expanded before transplantation. Following passaging and incubation, capabilities of proliferation and differentiation of MSCs decrease or even lose (24). Culture conditions such as serum source may impact on onset of replicative senescence. Compared with incubation in autologous serum, MSCs incubated with fetal bovine serum represent increase of telomere shorting as well as regulation of ROS and nitric oxide (25). Replicative senescence of MSCs is a continuous process that includes far reaching alterations in phenotype, differentiation potential, global gene expression patterns, and miRNA profiles (26). Expansion with basic fibroblast growth factor (bFGF) maintains MSC stemness by inhibiting cellular senescence through a PI3K/AKT-MDM2 pathway and by promoting proliferation, regulation of FGF receptors is involved in MSC self-renewal and inhibition of cellular senescence (27). Estrogen reduces aging of human MSCs by decreasing telomere shortening (28). Optimizing culture conditions, stem cell niches and systemic milieu are important strategies in preventing MSC senescence and improving therapeutic effectiveness of cell transplantation.

MSC senescence may be induced with long-term culture, ultraviolet radiation, X-ray, hyperoxia, ethanol, hydrogen peroxide, or *tert*-butylhydroperoxide in vitro. The senescent MSCs can be detected by examining population doublings, telomere shortening, SA-β-gal activity, expression of senescence-related genes such as $p16^{INK4a}$ (29–31). In this paper, we introduce the common methods to detect MSC senescence and investigate the effects of Wnt/β-catenin signaling on MSC aging.

2 Materials

2.1 Common Materials

2.1.1 Disposables

1. 25 cm^2 cell culture flasks (Corning, NY).
2. 35 × 10 mm cell culture dishes (Corning) or 6-well plates (Nunc, Denmark).
3. 22 × 22 mm coverslips for immunocytochemistry (Carl Roth GmbH, Karlsruhe, Germany) (see Note 1).
4. 15- and 50-mL conical centrifuge tubes (BD Falcon, San Jose, CA).
5. 1.5-mL microcentrifuge tubes (Eppendorf, Hamburg, Germany).
6. 5-mL, 12 × 75 mm FACS tubes (BD Falcon).
7. 75 × 25 mm microscope glass slides for fluorescence microscopy (Premiere, China).

2.1.2 Solutions

1. Growth medium for MSCs: Dulbecco's modified Eagle's medium (DMEM) supplemented with 10% fetal bovine serum (FBS) (Invitrogen, Carlsbad, CA).
2. 100x penicillin-streptomycin (10,000 U penicillin, 10 mg streptomycin/mL utilizing penicillin G and streptomycin sulfate in 0.85% saline) (Invitrogen).
3. Phosphate-buffered saline (PBS): 137 mM NaCl, 2.7 mM KCl, 4.3 mM Na_2HPO_4, and 1.4 mM KH_2PO_4 in deionized water (dH_2O), adjusted to pH 7.4 by titration with 1N NaOH.
4. Trypsin/ethylenediaminetetraacetic acid (EDTA) solution: 0.25% trypsin, 0.38 g/L (1 mM) EDTA·4Na in Hank's balanced salt solution (HBSS) (Invitrogen).

2.2 Induction of MSC Senescence

2.2.1 Bone Marrow-Derived Mesenchymal Stem Cells

1. MSCs used for the experiments: harvested from bone marrow of SD rats (12–14 weeks old) by adherent culture and subculture (see Notes 2 and 3). All the experiments described below are performed using MSCs of the third to fifth passage.
2. Old MSCs: harvested from bone marrow of old SD rats (64–68 weeks old) and used for detecting cellular senescence.
3. Young MSCs: harvested from bone marrow of young SD rats (3–4 weeks old) and used for analyzing cell cycle.

2.2.2 Serum of Rats

1. Old rat serum (ORS): purified with whole blood of old SD rats (64–68 weeks old) by clotting at 37°C and centrifugation (20).
2. Young rat serum (YRS): obtained from blood of young SD rats (3–4 weeks old) (20).

2.3 SA-β-gal Detection

Senescence associated β-galactosidase (SA-β-gal) is a biomarker associated with the senescent phenotype, SA-β-gal activity may be detected by histochemical staining in artificial substrate X-gal. SA-β-gal catalyzes hydrolysis of X-gal, which produces a blue color in senescent cells (30). The method to detect SA-β-gal is specific and convenient for assessing cellular senescence.

1. Senescence β-galactosidase staining kit (Genmed Scientifics, China) (see Note 4). SA-β-gal staining solution can be also prepared on the basis of followings.
   (a) Fixative solution: 3.7% (v/v) formaldehyde, or 4% (w/v) paraformaldehyde in PBS (see Note 5).
   (b) X-gal solution: 20 mg/mL X-gal stock solution in dimethylformamide (see Note 6).
   (c) 0.2 M citric acid/sodium phosphate buffer (pH 6.0): 36.85 mL of 0.1 M citric acid solution is mixed with 63.15 mL of 0.2 M sodium phosphate (dibasic) solution. Verify that the pH is 6.0, store at room temperature (see Note 7).
   (d) Potassium ferrocyanide solution: 500 mM potassium ferrocyanide in distilled water, store at 4°C (see Note 8).
   (e) Potassium ferricyanide solution: 500 mM potassium ferricyanide in distilled water, store at 4°C (see Note 9).
2. Mounting medium: 20% glycerol in PBS.
3. 37°C incubator (without CO_2).
4. Phase-contrast microscope (Nikon, Japan) or light microscope (Olympus, Japan).

2.4 ROS Assay

Excess ROS, the by-products of oxidative energy metabolism, can damage cells by peroxidizing lipids and disrupting structural proteins, enzymes and nucleic acids. The accumulated ROS damage may induce cellular senescence. Inversely, ability of the senescent cells against ROS stress deceases, resulting in increasing of ROS level. Generation of intracellular ROS is assessed with peroxide-sensitive fluorescent probe 2′,7′-dichlorofluorescein-diacetate (DCFH-DA). After permeating through cell membrane, DCFH-DA is hydrolyzed by intracellular esterases to DCFH. In the presence of ROS and DCFH is rapidly oxidized to highly fluorescent 2′,7′-dichlorofluorescein (DCF) (32).

1. 10 mM 2′,7′-dichlorodihydrofluorescein diacetate (DCFH-DA) (Molecular Probes) in PBS, store at −20°C.
2. 5 g/L Hoechst 33342 (Invitrogen, CA) in PBS, store at −20°C.
3. Fluorescence microscope (Olympus) or confocal laser scanning microscope (Carl Zeiss, Jena, Germany).
4. Flow cytometer (Calibur, BD Biosciences, USA) or Fluor spectrophotometer (Zenith 3100) (see Note 10).

2.5 Cell Cycle Assay

Cellular senescence is characterized by an irreversible growth arrest and certain altered functions. Cell cycle of senescent cells distributes abnormal, tends to stop in G_0/G_1 phase.

1. MSCs: harvested from young SD rats (3–4 weeks old) and old SD rats (64–68 weeks old) respectively.
2. Propidium iodide (PI) staining solution: 100 mg/L PI, 1% Triton X-100, 0.9 g/L NaCl, store at 4°C.
3. 10 g/L RNase-A in dH_2O, store at −20°C.
4. Fixative solution: 70% alcohol.
5. Disposable cell filters for flow cytometry.
6. Flow cytometer (Calibur, BD Biosciences).

2.6 Expression of Wnt-β-Catenin Signaling Molecules

Increasing studies demonstrated that an aged cell-extrinsic environment plays an important role in the aging of adult stem cells (18, 19). Recently, we have suggested that Wnt/β-catenin plays an important role on induction of MSC senescence (20). To determine the effects of ORS on Wnt/β-catenin signaling, we examine the expression of β-catenin and GSK-3β with immunofluorescent staining in this paper. β-catenin overexpression induces γ-H2A.X expression and causes p53 accumulation. A target gene of p53 protein, p21, can directly induce cellular aging.

1. Fixative solution: 4% PFA (Sigma, St. Louis, MO) in PBS, store at 4°C or prepare freshly prior to be used (see Note 5).
2. Permeabilizing solution: 0.5% Triton in PBS.
3. Blocking solution: 2% (w/v) bovine serum albumin (BSA) (Sigma) in PBS, or 5% (v/v) normal goat serum (Invitrogen) (see Note 11).
4. Antibody dilution buffer: 0.5% BSA in PBS (or commercially available from Beyotime Biotechnology).
5. Antibodies applied for immunofluorescence microscopy assessments are listed in Table 1.
6. 4′,6-Diamidino-2-phenylindole (DAPI, Sigma) staining solution, store at −20°C.
7. Mounting medium: 20% glycerol in PBS or SlowFade light (Invitrogen).
8. Fluorescence microscope (Olympus) or confocal laser scanning microscope (Zeiss LSM 510, Carl Zeiss).

2.7 Inhibition of Wnt-β-Catenin Signaling Pathway

2.7.1 RNA Interference for β-Catenin

siRNA oligonucleotides are synthesized by Genepharma Co. (Shanghai). The effective sequence used for the specific silencing of β-catenin is 5′-CACCTCCCAAGTCCTTTAT-3′ (si-β-catenin). The non-silencing control siRNA is an irrelevant siRNA with random nucleotide 5′-TTCTCCGAACGTGTCACGT-3′ and is not homologous to any sequences found in the gene bank.

Table 1
Antibodies for immunofluorescence staining

Primary antibody	Specificity	Source organism	Company[a]
Anti-GD2	Monoclonal IgG	Mouse	Imgenex
Anti-β-catenin	Polyclonal IgG	Rabbit	Cell Signaling
Anti-GSK-3β	Polyclonal IgG	Rabbit	Santa Cruz
Anti-γ-H2A.X	Polyclonal IgG	Rabbit	Santa Cruz
Anti-p53	Polyclonal IgG	Rabbit	Santa Cruz
Secondary antibody	**Label**	**Source organism**	**Company[a]**
Anti-rabbit	cy3-conjugated	Goat	ICN Cappel
Anti-mouse	FITC-conjugated	Goat	Dako

[a]Imgenex, USA; Cell Signaling, USA; Santa Cruz Biotechnology, Santa Cruz, CA; ICN Cappel, USA; Dako, USA

Table 2
siRNAs used to silence β-catenin gene

siRNA oligonucleotide	Sequence
Specific sequence of β-catenin	5′-CACCTCCCAAGTCCTTTAT-3′
Random nucleotides	5′-TTCTCCGAACGTGTCACGT-3′

1. siRNAs used for silencing β-catenin gene in MSCs are listed in Table 2.
2. HiPerFect transfection reagent (Qiagen, USA).
3. MSCs: isolated from SD rats aged 12–14 weeks and used for transfection and RNA interference (RNAi).
4. ORS: for induction of MSC aging, 20% ORS is supplemented in the growth medium.

2.7.2 Blocking of Wat/β-Catenin Signaling Pathway

1. Dickkopf-1 (DKK1) (R&D Systems, USA): 100 mg/mL stock solution, store at −20°C.
2. MSCs: the same as above.
3. ORS: the same as above.

2.8 RT-PCR Analysis for Expression of Wat/β-Catenin Signaling Molecule mRNAs

Wnt/β-catenin signaling is activated by the binding of Wnt ligands to the frizzled family of receptors. In the absence of Wnt ligands, β-catenin is phosphorylated by glycogen synthase kinase-3β (GSK-3β) and then degraded by the ubiquitin-proteasome system. When Wnt ligands bind to the frizzled receptors, GSK-3β activity is inhibited, and unphosphorylated β-catenin accumulates in the cytoplasm and translocates into the nucleus, where it promotes the

Table 3
Sequence of primers for PCR reactions

Target gene		Sequence
c-myc	Sense Antisense	5′-CCTACCCTCTCAACGACAGC-3′ 5′-CTCTAGCCTTTTGCCAGGAG-3′
p16^{INK4a}	Sense Antisense	5′-ACCAAACGC CCCGAACA-3′ 5′-GAGAGCTGCCACTTTGACGT-3′
p53	Sense Antisense	5′-ACCATGAGCGCTGCTCAGAT-3′ 5′-AGTTGCAAACCAGAC CTCAG-3′
p21	Sense Antisense	5′-TGAATGAAGGCTAAGGCAGAAGA-3′ 5′-AGGCAGACCAGCCTAACAGATT-3′
β-actin	Sense Antisense	5′-AAGAGAGGC ATCCTCACCCT-3′ 5′-TACATGGCTGGGGTGTTGAA-3′

transcription of a variety of the target genes such as c-myc (33). On the other hand, recent studies suggest that activated Wnt/β-catenin signaling can induce the DNA damage response. DNA damage response induces p16^{INK4a} expression (34, 35). The p16^{INK4a}, a tumor suppressor gene, is also an aging-induced gene that directly induces cellular aging (36, 37). In our experiments, the expression of c-myc, p16^{INK4a}, p53, and p21 mRNA in MSCs were detected.

1. RNA PCR kit (TaKaRa, Japan): store at −20°C.
2. TRIzol reagent (Invitrogen): store at 4°C.
3. PCR primers are listed in Table 3.
4. Agarose (TaKaRa).

2.9 Western Blot Analysis for Expression of GSK-3β, γ-H2A.X, and p53, and Translocation into the Nucleus of b-Catenin

1. RIPA lysis buffer: 50 mM Tris–HCl (pH 7.4), 150 mM NaCl, 1% Triton X-100, 0.5% NaDOD, and 0.1% SDS.
2. Cytoplasmic extraction reagents: 10 mM HEPES (pH 7.9), 10 mM KCl, 0.1 mM EDTA, 1.5 mM $MgCl_2$, 0.2% NP-40, 100 μM DTT, 50 μM PMSF (or commercially available from Pierce Chemical Company, USA).
3. Nuclear extraction reagents: 20 μM HEPES (pH 7.9), 10 mM KCl, 0.1 mM EDTA, 1.5 mM $MgCl_2$, 25% glycerol, 100 μM DTT, 50 μM PMSF (or commercially available from Pierce Chemical Company).
4. Enhanced BCA (Bicinchoninic acid) protein assay kit (Pierce Biotechnology, USA).
5. 30% acrylamide/*N*-*N*′-methylenediacrylamide (29/1) solution (Beyotime Biotechnology, Shanghai).
6. *N*,*N*,*N*′,*N*′-Tetra-methyl-ethylenediamine (TEMED, Beyotime Biotechnology), stored at 4°C (see Note 12).

Table 4
Antibodies for immunoblotting

Primary antibody	Specificity	Source organism	Company[a]
Anti-β-actin	Polyclonal IgG	Rabbit	Biovision
Anti-β-tubulin	Polyclonal IgG	Rabbit	Cell Signaling
Secondary antibody	**Specificity**	**Source organism**	**Company[a]**
Anti-rabbit	Horseradish peroxidase	Goat	Dako

[a]Biovision, USA; Cell Signaling, USA; Dako, USA

7. Ammonium persulfate (APS): 10% (w/v) solution in dH_2O, store at 4°C for a few days. It is better to prepare freshly prior to be used (see Note 13).
8. 0.1% (w/v) SDS in PBS.
9. Loading buffer (5×): 313 mM Tris–HCl (pH 6.8), 10 mM EDTA, 500 mM DTT, 10% (w/v) SDS, 50% (v/v) glycerol, 0.05% (w/v) bromophenol blue in dH_2O.
10. Transfer buffer: 25 mM Tris–HCl, 192 mM glycine, 20% methanol, in dH2O.
11. 3 MM® Whatmann filter paper.
12. Antibodies used for immunoblotting are listed in Tables 1 and 4.
13. Blocking buffer: 5% (w/v) nonfat powdered milk in TBST containing 0.1% Tween-20.
14. Enhanced chemiluminescence (ECL) reagents (Amersham Biosciences, USA).
15. Polyvinylidene difluoride (PVDF) membranes (Millipore, Bedford, MA).
16. Ponceau S staining solution: 0.1% (w/v) Ponceau S and 5% (w/v) acetic acid in dH_2O (or commercially available from Beyotime Biotechnology).
17. Photographic films: 5×7 in. Keda X-OMTA BT (Beyotime Biotechnology).
18. Prestained dual color protein molecular weight (MW) markers: Precision Plus Protein™ Standard, Dual Color (Beyotime Biotechnology).
19. Rinsing buffer: 0.05% Tween-20 in PBS (PBST) or Tris-buffered saline containing 0.05% Tween-20 (TBST) solution, storage at room temperature.
20. Running gel buffer: 1.5 M Tris buffer (pH 8.8).
21. Stacking gel buffer: 1.0 M Tris buffer (pH 6.8).
22. Electrophoresis apparatus: Mini-PROTEAN Tetra cell electrophoresis system (Bio-Rad, Hercules, CA).

3 Methods

3.1 Purification of Serum

1. Whole blood is collected from the anesthetized young SD rats (3–4 weeks old) or old SD rats (64–68 weeks old) via the abdominal aorta.
2. Blood is clotted at 37°C for 4 h.
3. Serum is isolated by centrifugation (1,200 × *g*, for 10 min).
4. The supernate is collected and then centrifugation is repeated once.
5. The supernate is collected and stored at –20°C.

3.2 Detection of MSC Senescence

3.2.1 Isolation of MSCs

1. The femurs and tibias of SD rats aged 12–14 weeks (adult rats) or 64–68 weeks (old rats) are removed.
2. Bone marrow is harvested by flushing with PBS containing 100 U/mL heparin.
3. Centrifuge the cells at 200 × *g* for 10 min.
4. Resuspend the cells in DMEM supplemented with 10% FBS.
5. Plate the cells in a 25 cm^2 plastic flask to allow the MSCs to adhere.
6. Change the medium at day 3 after incubation to discard non-adherent cells (see Note 14).
7. Replace the medium completely every 3 days.
8. Harvest MSCs with treatment in 0.25% trypsin/1 mM EDTA solution when the cells grow to 70~80% confluence, and then seed the cells in flasks (see Note 3).
9. MSCs are identified by expression of CD 105, CD 166, Thy-1, and GD2.

3.2.2 Induction of MSC Senescence with ORS

1. Incubate the cells with DMEM containing 20% ORS for 36 h (see Note 15).
2. For inhibition of Wnt/β-catenin signaling pathway, 100 ng/mL DKK1 is added in DMEM containing 20% ORS and then the cells are incubated for 36 h.
3. For siRNA-mediated inhibition of Wnt/β-catenin signaling (RNAi of β-catenin), the cells are transfected with HiPerFect Transfection Reagent containing si-β-catenin for 12 h (see Note 16). Then, 20% ORS is added in the medium and the cells are incubated for 36 h.

3.3 SA-β-Gal Staining

The working solution is prepared according instructions of senescence β-galactosidase staining kit and based on the number of samples. If the kit is not available, the staining solution is prepared by mixing 1 mg/mL of X-gal, 40 mM citric acid/sodium phosphate

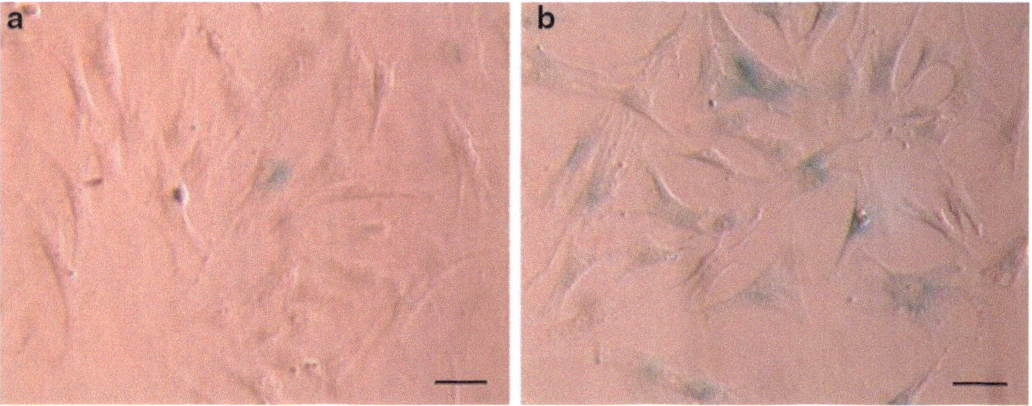

Fig. 1 Expression of SA-β-gal in MSCs treated with ORS. (**a**) MSCs were incubated with DMEM containing 20% YRS for 36 h. A few of the cells are positive for SA-β-gal staining. (**b**) MSCs were incubated with DMEM containing 20% ORS for 36 h. Some cells display a large and flattened morphology. SA-β-gal-positive cells increased. Bars = 20 μm

buffer (pH 6.0), 5 mM potassium ferricyanide, 5 mM potassium ferrocyanide, 150 mM NaCl, and 2 mM $MgCl_2$ (see Note 17).

1. The cells are incubated in 35 × 10 mm culture dishes or 6-well plates. When the cells grew to 70~80% confluence, aspirate the medium and rinse the cells twice with PBS.
2. Add 1 mL of fresh 4% paraformaldehyde (per dish or per well), fix the cells at room temperature for 5 min.
3. Remove the fixing solution and rinse the fixed cells with PBS for three times.
4. Add 1 mL fresh staining solution (per dish or per well) and incubate the cells avoiding of light at 37°C (not in CO_2 incubator) from 2 h to overnight (see Note 18).
5. Remove the staining solution, and then wash the cells twice with PBS and store cells in PBS. For long-term storage at 4°C, overlay the cells with mounting medium.
6. View the positive blue cells with a light microscope after staining for 12 h (see Note 19). An example of SA-β-gal staining of the senescent MSCs induced with ORS is shown in Fig. 1.
7. Count the positive cells and calculate ratio of the positive cells to total cells.

3.4 Fluorescence Analysis of ROS

1. MSCs are grown in 6-well plates or on coverslips and then treated with ORS for 36 h.
2. After aspirating the medium, the cells are rinsed twice with PBS.
3. DCFH-DA stock solution is diluted to a final concentration (10 μM) with serum-free DMEM prior to be used (see Note 20).

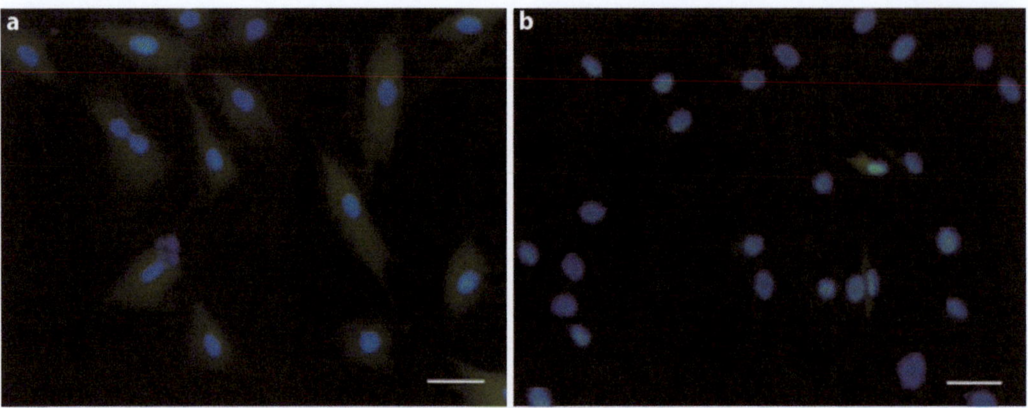

Fig. 2 ROS generation of MSCs. (**a**) The cells were incubated with DMEM containing 20% ORS for 36 h. The most cells are positive for ROS assay. (**b**) The cells were treated with DMEM containing 20% ORS and 100 ng/mL DDK1 (an inhibitor of Wat/β-catenin signaling pathway) for 36 h. Only a few of the cells are positive. Bars = 20 μm

4. Add 10 μM DCFH-DA to 6-well plates and incubate the cells at 37°C for 15 min (see Note 21).
5. After washing for three times with PBS, the nuclei are stained with Hoechst 33342 (1:1,000) (see Notes 22 and 23).
6. DCF fluorescence is viewed using a fluorescence microscope (Olympus Corporation), the level of ROS is evaluated according to intensity of DCF fluorescence by a flow cytometer or a fluorospectrophotometer (see Note 24). An example of ROS assay of senescent MSCs induced with ORS is showed in Fig. 2.

3.5 Cell Cycle Assay

1. MSCs are obtained from young and old rats respectively. The cells are collected by trypsinization. After washing twice with PBS, cell density is adjusted to 1×10^6 cell/mL.
2. Centrifuge the cells at 300×*g* for 10 min. After discarding the supernate, the cells are fixed with 70% alcohol at 4°C overnight.
3. Rinse the fixed cells for three times with PBS, the cells are resuspended with 0.4 mL PBS.
4. Add 50 mg/L RNase-A, incubate the cells in a thermostat waterbath at 37°C for 30 min.
5. After filtrating with disposable cell filters, add PI staining solution into cell suspensions. The cells are incubated protecting from light at 4°C for 30 min (see Note 25).
6. Cell cycle is analyzed by a flow cytometer. An example for the cell cycles of the MSCs obtained from young and old rats is shown in Fig. 3.

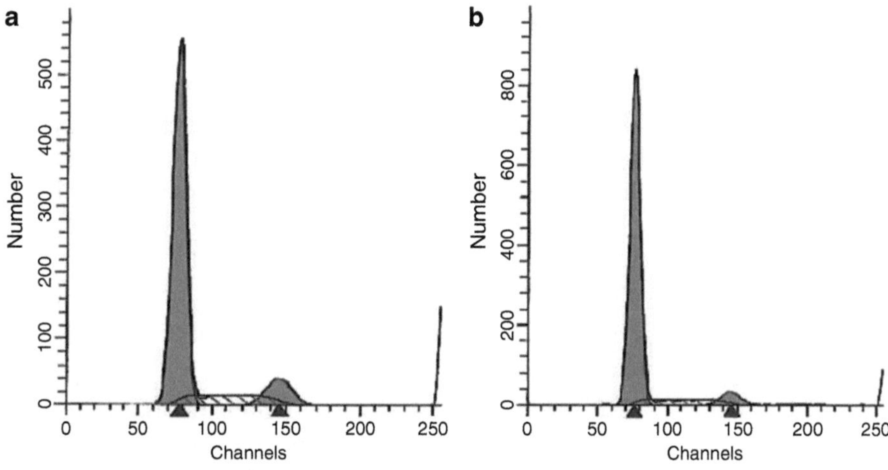

Fig. 3 The cell cycle of aging MSCs. (**a**) The MSCs isolated from young rats (3–4 weeks old). There are 11.25% cells in S phase. (**b**) The MSCs isolated from old rats (64–68 weeks old). The cell cycles of the most of the cells are stop in G_0/G_1 phase. (*Filled triangle*) indicates G_0/G_1 phase cells and G_2/M phase cells respectively. Biases indicate S phase cells

3.6 Rescence Analysis

To examine β-catenin expression in MSCs, primary MSCs are isolated from SD rats aged 12–14 and 64–68 weeks as previously described (20, 38). GD2 is used as a marker of MSCs.

1. Seed the cells in 35 × 10 mm dishes or on coverslips, incubate in DMEM without serum for 24 h.
2. The cells are fixed with 4% (w/v) paraformaldehyde at room temperature for 10 min.
3. After washing with PBS, the cells are blocked with 2% BSA for 30 min (see Note 11).
4. The cells are incubated with rabbit anti-rat β-catenin (1:200) and mouse anti-rat GD2 (1:200) at 37°C for 1 h or at 4°C overnight.
5. After washing, the cells are incubated with goat anti-rabbit (cy3)-conjugated antibodies (1:300) and goat anti-mouse FITC-conjugated antibodies (1:300) in the dark at 37°C for 30 min.
6. The nuclei are counterstained with DAPI (1:1,000).

For evaluating the effect of ORS on the expression of β-catenin, GSK-3β, γ-H2A.X and p53, MSCs from SD rats aged 12–14 weeks are used in the experiments.

1. The cells are seeded on coverslips in 35 × 10 mm dishes and incubated until the cells are grown to submonolayer.
2. After being fixed and blocked, the cells are incubated with rabbit anti-rat β-catenin (1:200), rabbit anti-rat GSK-3β (1:200), rabbit anti-rat γ-H2A.X (1:100), or rabbit anti-rat p53 (1:200) polyclonal antibodies.

3. After washing with PBS, the cells are incubated with goat anti-rabbit (cy3)-conjugated antibodies or FITC-conjugated antibodies (1:300).
4. DAPI (1:1,000) was used to visualize nuclei. After washing and being mounted on glass slides, the cells were examined using a fluorescence microscope or a confocal laser scanning microscope.

3.7 RT-PCR Analysis

1. Total RNA of the cells is extracted according to the protocol provided by the manufacturer.
2. The mRNA levels are normalized by using gapdh as a housekeeping gene.
3. PCR reactions are performed under the following conditions: 1 cycle at 45°C for 30 min and 94°C for 5 min, followed by 35 cycles at 94°C for 30 s, 60°C for 30 s and 72°C for 1 kb/min, and 1 cycle at 72°C for 5 min (23).
4. PCR products were size-fractionated with 1.5% agarose (TaKaRa) gel electrophoresis (22, 23).
5. The images are analyzed by a software (Furi Science & Technology).

3.8 Western Blot Analysis

To assay the β-catenin protein, cytoplasmic and nuclear proteins from the cultured cells are prepared using cytoplasmic and nuclear extraction reagents respectively. β-actin and β-tubulin are used as the internal control for the cytoplasmic and nuclear proteins respectively. To assay expression of GSK-3β, γ-H2A.X, and p53, the total cellular protein is extracted using the following methods.

3.8.1 Preparation of Samples

1. For western blot analysis, MSCs are incubated in 6-well plates or 25 cm^2 plastic flasks.
2. Following treatment with ORS to induce cellular senescence, MSCs are collected in Eppendorf (EP) tubes after trypsinization and washed once with ice-cold (4°C) PBS by centrifugation (300–500 × *g*, 5 min) at 4°C.
3. Supernatant is discarded and pellets are lysed in 30–50 μL of RIPA lysis buffer (see Note 26).
4. The lysates are ultracentrifuged (12,000 × *g*, 5 min) at 4°C and then supernatant is transferred to new EP tubes (see Notes 27 and 28).
5. Concentration of protein in the samples is determined by BCA protein assay, following the manufacturer's instructions.
6. 4–6 μL of 5× loading buffer is added to 20–80 μg of proteins from each lysate, the final volume is adjusted to 20–30 μL with dH_2O.
7. The samples are boiled for 5 min at 100°C. After cooling to room temperature, the samples are ready for separation by sodium dodecylsulfate-polyacrylamide gel electrophoresis (SDS-PAGE).

3.8.2 SDS-PAGE

1. 12.5% running gel is prepared by mixing 1.75 mL of 30% acrylamide:*N*-*N*′-methylenediacrylamide (29/1) solution, 1.25 mL of running gel buffer (pH 8.8), 50 μL of 10% SDS, 2 μL of TEMED, 50 μL of 10% APS, and 1.9 mL of dH_2O.
2. To avoid polymerization in tube, APS should be added as the last component of the gel solution. After addition of APS, the running gel should be immediately poured into preassembled glasses up to approximately two thirds of their height.
3. Overlay gel with 1~5 mL dH_2O to ensure a flat surface and exclude air, and polymerize the gel at room temperature for 20~40 min.
4. Pour off dH_2O and wash the top of the gel with new dH_2O to scour off unpolymerized acrylamide, and suck dry dH_2O with 3 MM filter paper.
5. 5% stacking gel is prepared by mixing 0.5 mL of 30% acrylamide/bisacrylamide (29/1) solution, 0.38 mL of stacking gel buffer (pH 6.8), 30 μL of 10% SDS, 3 μL of TEMED, 30 μL of APS, and 2.1 mL of dH_2O. APS should be added as the last component of the gel solution.
6. Rapidly following the addition of APS, the stacking gel is poured on top of the polymerized running gel and the comb for the creation of wells is inserted, and place 15~30 min until the stacking gel polymerized.
7. Once the stacking gel is polymerized, the comb is gently removed and the gel unit is moved to the cell for separation. Upper and lower chambers of the gel unit are filled with migration buffer (1×), which is used also to gently wash sample wells.
8. Each well is loaded with a sample prepared as described above. One well is reserved for protein MW markers.
9. The gel unit is connected to power supply and the proteins are separated in constant-field mode. The run should be arrested when MW markers are separated throughout entire height of the gel.

3.8.3 Tank Electrotransfer

1. The running gel is immersed in transfer buffer for 30 min.
2. One PVDF membrane of the same size as the running gel is immersed successively in methanol for 20 s, dH_2O for 5 min, and transfer buffer for 12 min.
3. Six sheets of 3 MM filter paper of the same size as the running gel are soaked with transfer buffer.
4. The transfer "sandwich" is prepared by overlaying the following components: three sheets of filter paper, the PVDF membrane, the running gel, three sheets of filter paper according to the sequence from anode to cathode.

5. The "sandwich" is placed in a transfer cassette, which subsequently inserted into electrophoresis apparatus, the transfer is accomplished in ice-bath at the constant voltage of 100 V for 1.5 h.
6. After the transfer completed, the colored MW markers should be visible on the PVDF membrane, and the bound proteins can be visualized by a rapid incubation with Ponceau S staining solution.
7. Ponceau S should be removed by rinsing the membrane with PBS or dH_2O prior to blocking.

3.8.4 Immunoblotting

1. To block unspecific binding sites, the membranes are blocked on a rocking platform in blocking buffer at room temperature for 45 min.
2. The PVDF membranes are incubated with the primary antibodies of rabbit anti-rat β-catenin, GSK-3β, γ-H2A.X, p53, and β-actin or β-tubulin (1:200) at 4°C overnight or at room temperature for 2 h.
3. After being washed three times with rinsing buffer, the membranes are incubated with horseradish peroxidase (HRP)-conjugated goat anti-rabbit IgG (1:2,000) at room temperature for 1 h.
4. After being again washed with rinsing buffer, the membrane is overlaid with ECL reagents and incubated at room temperature for 1 min. β-actin and β-tubulin are used as the internal control to normalize the loading materials.
5. The membrane is moved into an X-ray film cassette, the side of the proteins of the membrane is upwards, a photographic film is exposed to the proteins on the membrane for 3–4 min in a dark room.
6. The exposed films are developed in a common tabletop film processor, and then photographed and analyzed.

4 Notes

1. In order to facilitate cell adhesion and growth, it is required that the coverslips are immersed in the diluted hydrochloric acid or other acid solution for 12 h. After washing sufficiently by tap water and then by purifying deionized water for three times, the coverslips are sterilized by a high pressure steam sterilizer before use.
2. MSCs harvested from bone marrow of SD rats aged 12–14 weeks are appropriate for detection of the senescence-associated changes induced by ORS. MSCs from bone marrow may be isolated by adherent culture.

3. Subculture of MSCs is performed when the cells are grown to 70~80% confluence, preventing cellular senescence caused by excess density or passage.
4. The senescence β-galactosidase staining kit is used to detect the level of senescence-associated β-galactosidase (SA-β-gal) activity in senescent cells or tissue. The kit contains β-gal staining fixative solution 100 mL, X-gal solution 5 mL, β-gal staining solution A 1 mL, β-gal staining solution B 1 mL, and β-gal staining solution C 100 mL. According to the manufacturer's instructions, store X-gal solution should be protected from light at -20°C, and other components of the kit are stored at 4°C.
5. For preparation of 3.7% formaldehyde, 1 mL of 37% formaldehyde is added to 9 mL of PBS. The solution is freshly prepared for each experiment. For preparation of 4% (w/v) paraformaldehyde, 0.4 g paraformaldehyde is dissolved in 10 mL of PBS. After adding one drop of 1N NaOH, the solution is heated to 65–70°C until it becomes clear. Then, it is cooled at room temperature. It is freshly prepared or store at 4°C for a few days.
6. In preparation of X-gal solution, 20 mg/mL X-gal is dissolved in DMF in dark-colored or aluminum foil-wrapped glass vials or similar containers to protect from light. The solution may be stored at –20°C for a few days.
7. For preparation of 0.1 M citric acid solution, citric acid monohydrate ($C_6H_8O_7{\cdot}H_2O$) is dissolved in distilled water. For preparation of 0.2 M sodium phosphate solution, sodium dibasic phosphate (Na_2HPO_4) or sodium dibasic phosphate dehydrate ($Na_2HPO_4{\cdot}H_2O$) is dissolved in distilled water. The solutions may be kept at room temperature for several months.
8. Potassium ferricyanide is dissolved at 500 mM in distilled water and stored in a tube covered with aluminum foil to protect from light.
9. 500 mM potassium ferrocyanide solution is prepared in distilled water and stored in a tube covered with aluminum foil to protect from light.
10. DCFH-DA is applied in sensitive and rapid quantitation of ROS in response to oxidative metabolism. To quantify the ROS level, intensity of DCF fluorescence in the cells may be evaluated by flow cytometry at an excitation wavelength of 488 nm and an emission wavelength of 525 nm or fluorospectrophotometry.
11. If the background of immunostaining is high, which is uncommon in senescent cells, the staining may be improved by blocking with 5% goat serum or 5% nonfat milk in PBS, or diluting the primary and/or secondary antibodies.

12. TEMED is toxic by inhalation. It is better to store at 4°C. It should be always handled in a fume hood.
13. 10% APS stock solution (in dH_2O) is unstable. It is better to prepare it freshly. Storage at 4°C is possible, but it never exceed 1 week. The use of old APS solution may result in incomplete polymerization of the acrylamide gel.
14. It is important to remove non-adherent cells thoroughly for harvesting pure MSCs.
15. Duration of induction is dependent to concentration of ORS and response of the cells to inducible factors in ORS. If senescence-associated changes in the cells is not observed after induction, concentration of ORS should be increased, duration of induction may prolong.
16. After the time required for the siRNA-mediated downregulation of the target protein, the desired stimuli can be administered to the cells in which the Wnt/β-catenin signaling pathway has been suppressed by genetic techniques.
17. When fresh staining solution is prepared, please use polypropylene or glass container. Polystyrene container is not suitable. However, the staining procedure may be done in polystyrene container, such as in 6-well plates.
18. Remember that the cells are not incubated in CO_2 incubator. Because SA-β-gal staining is dependent on specific pH of the staining solution, CO_2 may change the pH of staining solution. Some positive cells (blue color) are detectable within 2 h under a phase-contrast microscope, but the staining is generally maximal at 12–16 h after incubation.
19. If the cells are incubated for 12 h or overnight, the culture dishes or 6-well plates should be overlayed with parafilm or preservative film to avoid the solution evaporating.
20. To test adequate concentration of DCFH-DA, it is better to use Rosup for positive control.
21. During incubation with DCFH-DA staining solution, culture dishes or tubes may be shaken gently at the interval of 3~5 min to let the probe to permeate into the cells.
22. Residual DCFH-DA out of the cells should be removed thoroughly with washing in PBS to less background.
23. 5 g/L Hoechst 33342 store solution is diluted with DMEM into 5 mg/L working solution.
24. Besides examination with fluorescence microscope or confocal laser scanning microscope, intensity of DCF fluorescence in the suspended cells may be analyzed with flow-cytometry or fluorospectrophotometry. In latter cases, it is unnecessary to stain nuclei with Hoechst 33342.

25. Duration of PI staining is too long to affect analysis of the results. Therefore, the cell cycles should be analyzed immediately by a flow cytometer after PI staining. Moreover, PI is a carcinogenic substance, avoiding touch directly.
26. For extraction of cytoplasmic proteins, the pellets are lysed in 30–50 μL of cytoplasmic extraction reagents after centrifugation.
27. In extraction of cytoplasmic proteins, the lysates swell on ice for 5 min. After centrifugation (500 × *g*, 5 min) at 4°C, supernatant containing cytoplasmic proteins is collected.
28. For extraction of nuclear proteins, the pellets are lysed in 30–50 μL of nuclear extraction reagents. After reaction on ice for 5 min, the lysates are ultracentrifuged (12,000 × *g*, 5 min) at 4°C and supernatant containing nuclear proteins is collected.

Acknowledgments

This work was supported by National Natural Science Foundation of China (30470883, 81270200 and 30971674), Scientific Research Foundation of State Education Commission (200802460044), and The Foundation of Science and Technology Commission of Shanghai Municipality (11540702500).

References

1. Campisi J, d'Adda di Fagagna F (2007) Cellular senescence: when bad things happen to good cells. Nat Rev Mol Cell Biol 8:729–740
2. Sharpless NE, DePinho RA (2007) How stem cells age and why this makes us grow old. Nat Rev Mol Cell Biol 8:703–713
3. Wallenfang MR (2007) Aging within the stem cell niche. Dev Cell 13:603–604
4. Pan L, Chen S, Weng C et al (2007) Stem cell aging is controlled both intrinsically and extrinsically in the Drosophila ovary. Cell Stem Cell 1:458–469
5. Carlson ME, Conboy IM (2007) Loss of stem cell regenerative capacity within aged niches. Aging Cell 6:371–382
6. Kovacic JC, Moreno P, Hachinski V et al (2011) Cellular senescence, vascular disease, and aging: part 1 of a 2-part review. Circulation 123:1650–1660
7. Caplan AI, Bruder SP (2001) Mesenchymal stem cells: building blocks for molecular medicine in the 21st century. Trends Mol Med 7:259–264
8. Książek K (2009) A comprehensive review on mesenchymal stem cell growth and senescence. Rejuvenation Res 12:105–116
9. Stolzing A, Jones E, McGonagle D et al (2008) Age-related changes in human bone marrow-derived mesenchymal stem cells: consequences for cell therapies. Mech Ageing Dev 129: 163–173
10. Zhou S, Greenberger JS, Epperly MW et al (2008) Age-related intrinsic changes in human bone-marrow-derived mesenchymal stem cells and their differentiation to osteoblasts. Aging Cell 7:335–343
11. Sethe S, Scutt A, Stolzing A (2006) Aging of mesenchymal stem cells. Ageing Res Rev 5: 91–116
12. Gerland LM, Peyrol S, Lallemand C et al (2003) Association of increased autophagic inclusions labeled for beta-galactosidase with fibroblastic aging. Exp Gerontol 38:887–895
13. Muller M (2009) Cellular senescence: molecular mechanisms, in vivo significance, and redox considerations. Antioxid Redox Signal 11:59–98

14. Shibata KR, Aoyama T, Shima Y et al (2007) Expression of the p16INK4A gene is associated closely with senescence of human mesenchymal stem cells and is potentially silenced by DNA methylation during in vitro expansion. Stem Cells 25:2371–2382
15. Sherman MH, Bassing CH, Teitell MA (2011) Regulation of cell differentiation by the DNA damage response. Trends Cell Biol 21:312–319
16. So AY, Jung JW, Lee S et al (2011) DNA methyltransferase controls stem cell aging by regulating BMI1 and EZH2 through microRNAs. PLoS One 6:e19503
17. Kim JS, Kim EJ, Kim HJ et al (2011) Proteomic and metabolomic analysis of H_2O_2-induced premature senescent human mesenchymal stem cells. Exp Gerontol 46:500–510
18. Conboy IM, Conboy MJ, Wagers AJ et al (2005) Rejuvenation of aged progenitor cells by exposure to a young systemic environment. Nature 433:760–764
19. Villeda SA, Luo J, Mosher KI et al (2011) The ageing systemic milieu negatively regulates neurogenesis and cognitive function. Nature 477:90–94
20. Zhang DY, Wang HJ, Tan YZ (2011) Wnt/b-Catenin signaling induces the aging of mesenchymal stem cells through the DNA damage response and the p53/p21 pathway. PLoS One 6:e21397
21. Sensebe L, Krampera M, Schrezenmeier H et al (2010) Mesenchymal stem cells for clinical application. Vox Sang 98:93–107
22. Guo HD, Wang HJ, Tan YZ et al (2011) Transplantation of marrow-derived stem cells carried in fibrin improves cardiac function after myocardial infarction. Tissue Eng Part A 17:45–58
23. Wu JH, Wang HJ, Tan YZ et al (2012) Characterization of rat very small embryonic-like stem cells and cardiac repair after cell transplantation for myocardial infarction. Stem Cells Dev 21(8):1367–1379. Epub 2011 Oct 27
24. Bonab MM, Alimoghaddam K, Talebian F et al (2006) Aging of mesenchymal stem cell in vitro. BMC Cell Biol 7:14
25. Duggal S, Brinchmann JE (2011) Importance of serum source for the in vitro replicative senescence of human bone marrow derived mesenchymal stem cells. J Cell Physiol 226:2908–2915
26. Wagner W, Horn P, Castoldi M et al (2008) Replicative senescence of mesenchymal stem cells: a continuous and organized process. PLoS One 3:e2213
27. Coutu DL, Francois M, Galipeau J (2011) Inhibition of cellular senescence by developmentally regulated FGF receptors in mesenchymal stem cells. Blood 117:6801–6812
28. Breu A, Sprinzing B, Merkl K et al (2011) Estrogen reduces cellular aging in human mesenchymal stem cells and chondrocytes. J Orthop Res 29:1563–1571
29. Itahana K, Campisi J, Dimri GP (2007) Methods to detect biomarkers of cellular senescence: the senescence-associated beta-galactosidase assay. Methods Mol Biol 371:21–31
30. Debacq-Chainiaux F, Erusalimsky JD, Campisi J et al (2009) Protocols to detect senescence-associated beta-galactosidase (SA-betagal) activity, a biomarker of senescent cells in culture and in vivo. Nat Protoc 4:1798–1806
31. Wagner W, Bork S, Lepperdinger G et al (2010) How to track cellular aging of mesenchymal stromal cells? Aging 2:224–230
32. LeBel CP, Ischiropoulos H, Bondy SC (1992) Evaluation of the probe 2′,7′-dichlorofluorescin as an indicator of reactive oxygen species formation and oxidative stress. Chem Res Toxicol 5:227–231
33. Clevers H (2006) Wnt/beta-catenin signaling in development and disease. Cell 127: 469–480
34. Collins CJ, Sedivy JM (2003) Involvement of the INK4a/Arf gene locus in senescence. Aging Cell 2:145–150
35. Kosar M, Bartkova J, Hubackova S et al (2011) Senescence-associated heterochromatin foci are dispensable for cellular senescence, occur in a cell type- and insult-dependent manner and follow expression of p16 (ink4a). Cell Cycle 10:457–468
36. Cànepa ET, Scassa ME, Ceruti JM et al (2007) INK4 proteins, a family of mammalian CDK inhibitors with novel biological functions. IUBMB Life 59:419–426
37. Janzen V, Forkert R, Fleming HE et al (2006) Stem-cell ageing modified by the cyclin-dependent kinase inhibitor p16INK4a. Nature 443:421–426
38. Cui XJ, Xie H, Wang HJ et al (2010) Transplantation of mesenchymal stem cells with self-assembling polypeptide scaffolds is conducive to treating myocardial infarction in rats. Tohoku J Exp Med 222:281–289

Chapter 10

Intra-femoral Injection of Human Mesenchymal Stem Cells

Sindhu T. Mohanty and Ilaria Bellantuono

Abstract

In vivo transplantation of putative populations of hematopoietic stem cells (HSC) and assessment of their engraftment is considered the golden standard to assess their quality and degree of stemness. Transplantation is usually carried out by intravenous injection in murine models and assessment of engraftment is performed by monitoring the number and type of mature blood cells produced by the donor cells in time. In contrast intravenous injection of mesenchymal stem cells (MSC), the multipotent stem cells present in bone marrow and capable of differentiating to osteoblasts, chondrocytes and adipocytes, has not been successful. This is due to limited or absent engraftment levels. Here, we describe the use of intra-femoral injection as an improved method to assess MSC engraftment to bone and bone marrow and their quality.

Keywords Mesenchymal stem cells, Marrow stromal cells, Intra-femoral injection, Gene marking, Lentiviral transduction, Enhanced green fluorescent protein

1 Introduction

Integral to the definition of stem cells is their ability to regenerate the tissue in which they reside through their ability to self-renew and differentiate. Stem cells are thought to play a role in maintenance and repair of tissues. Studies in murine models have highlighted the importance of testing the regenerative capacity of stem cells by transplantation in the context of ageing to determine the changes they undergo with in vivo or during expansion in culture and how this impact on tissue homeostasis. For example HSC from 22–24-month-old mice have been shown decreased engraftment ability following transplantation compared to younger mice and a skewed regeneration of the myeloid lineage at the expense of the lymphoid lineage (1).

Mesenchymal stem cells (MSC) reside in bone marrow and are able to differentiate to osteoblasts, adipocytes, chondrocytes and hematopoietic supporting stroma (2, 3). They have an important role in repair and maintenance of bone and the bone marrow microenvironment. Moreover, loss of proliferation and differentiation ability has been reported with age in vitro (4). However, robust

Kursad Turksen (ed.), *Stem Cells and Aging: Methods and Protocols*, Methods in Molecular Biology, vol. 976, DOI 10.1007/978-1-62703-317-6_10,

evidence of reduced regenerative capacity in vivo is scant. This is due to the limitation of the current transplantation methodologies available. The best in vivo assay to assess MSC regenerative capacity and quality is the ectotopic bone formation assay. This consists of seeding MSC in an appropriate porous scaffold and implanting this under the skin of a mouse. Five weeks later an ossicle is formed with areas of chondrogenesis and host hematopoiesis (5), allowing assessment of their ability to undergo osteogenic, adipogenic, and chondrogenic differentiation. However, this assay does not test the ability of MSC to regenerate and contribute to tissue maintenance and repair in the appropriate environment. To test this we describe a method where labelled human MSC are injected directly in the femur of immunodeficient mice and their engraftment is assessed 5 weeks later.

2 Materials

The culture of human MSC and the preparation of all reagents require sterile conditions and are carried out in a class II biological safety cabinet. The intra-femoral injection is a regulated procedure and requires a licence according to the regulations for animal procedures imposed by the country where the procedure is taking place. Lentiviral work will require appropriate genetic modified organism risk assessment and approval according to the regulations of the country where the work takes place. Follow waste disposal regulations when disposing of materials.

2.1 Cell Preparation

1. Human mesenchymal stem cell medium: Dulbecco's modified eagle medium (DMEM, Life technologies, Paisley, UK) containing 10% fetal calf serum (Hyclone, Fisher Scientific, Loughborough, UK). Store at 4°C.
2. MSC isolated from human bone marrow and expanded in culture (see Note 1).
3. Lentiviral particle containing a vector expressing enhanced green fluorescent protein (eGFP) at a concentration of 10^6 vp/ml minimum (see Note 2).
4. 0.05% Trypsin–0.53 mM EDTA (Ethylenediaminetetraacetic acid) (Life technologies) store at 4°C.
5. Sterile phosphate buffered saline (PBS).
6. Sterile Dulbecco's modified eagle medium (DMEM). Store at 4°C.

2.2 Surgical Components and Reagents

1. A pair of sterile scissors, forceps, artery forceps, scalpel blade, scalpel holder, and surgical drapes (see Note 3).
2. Cork sheet.

3. Surgical electric shaver.
4. Three 1 ml 27 g × ½ inch syringes with needles (see Note 4).
5. Hamilton syringe and removable needles (RN) (Hamilton, Bonaduz, Switzerland) (see Note 5).
6. Needles 25 g to position the mouse and keep it steady.
7. Blue monofilament adsorbable suture 45 cm (Ethicon, Edinburgh, UK).
8. Ketaset and Rompun (see Note 6).
9. Bone wax (Ethicon).
10. Sterile ethanol wipes.
11. Sterile distilled water and PBS.
12. Temperature controlled incubators set at 37°C.
13. NOD/LtSz-Prkdc*scid* (NOD/SCID) mice 5–6 weeks old.

2.3 Immuno-Staining Reagents

1. 10% neutral buffered formalin: Weigh 8 g Sodium dihydrogen orthophosphate dehydrate, 13 g Disodium hydrogen orthophosphate dehydrate. Add 200 ml concentrated formaldehyde (i.e., 37–41%) and 200 ml warm tap water (helps to dissolve buffers). Mix to dissolve buffers and top up with tap or distilled water to a final volume of 2 l. Store at room temperature (see Note 7).
2. Neutral EDTA: Weigh 250 g Ethylenediaminetetraacetic acid (disodium salt), 25 g Sodium hydroxide and mix with 1,750 ml Distilled water. The solution will be cloudy until the addition of sodium hydroxide which will also neutralize it to pH 7. Mix well by stirring and store at room temperature.
3. SuperFrost Plus slides (VWR International Ltd, Lutterworth, UK).
4. Xylene (VWR International Ltd).
5. 99% absolute industrial methylated spirit (ethanol) (Thermo Fisher UK Ltd, Loughborough, UK).
6. Leica decalcifier II (Leica Microsystems, Milton Keynes, UK).
7. Leica RM2265 rotary microtome (Leica Microsystems).
8. Leica TP1020 carousel tissue processor (Leica Microsystems).
9. Histology wax (Leica Microsystems).
10. 3% hydrogen peroxide: add 3 ml of hydrogen peroxide (VWR International Ltd) in 97 ml of PBS.
11. PBS.
12. Normal Goat serum (Dako UK Ltd, Ely, UK).
13. Anti-GFP, rabbit IgG fraction (polyclonal) (Life Technologies).
14. Goat anti-rabbit horseradish peroxidase (HRP) (Insight Biotechnology Ltd, Wembley, UK).

15. Vector staining kit (Vector laboratories Ltd, Peterborough, UK).
16. Gill's hematoxylin stain (Merck chemicals Ltd, Nottingham, UK).
17. Di-*n*-Butyl Phthalate in Xylene (DPX) (VWR International Ltd).

3 Methods

3.1 Labelling of hMSC to Express eGFP by Lentiviral Transduction

All procedures are carried out in a Class II biological safety cabinet.

1. Human MSCs previously isolated and expanded in MSC medium and devoid of any hematopoietic contamination are plated at a density of 10,000/cm^2 in MSC medium.
2. The day after, MSC are incubated with MSC medium containing viral particles at a multiplicity of infection of 40–60 and left at 37°C in 5% carbon dioxide (CO_2) in air for 8 h.
3. The media is then removed and fresh MSC medium is added to the cells which are further incubated for 5 days at 37°C in 5% CO_2.
4. Cells are washed once with PBS (10 ml/T25) and incubated with Trypsin/EDTA (1 ml/25 cm^2) for 2 min at 37°C in 5% CO_2 in air (see Note 8).
5. Cells are then detached from the flask and harvested using MSC medium (10 ml/T25 flask).
6. An aliquot of the cells is removed to determine the transduction efficiency by assessing the expression of eGFP by fluorescent activated cell sorting (FACS).
7. Human MSC showing greater than 90% eGFP expression are suitable for transplantation and are replated in MSC medium for expansion until the correct number of cells required for transplantation is obtained.

3.2 Preparation of hMSC for Injection

1. Use human MSC culture which are 80–90% confluent for transplant.
2. Detach the cells from the flask as described in Subheading 3.1, steps 4 and 5.
3. Centrifuge at 800 × *g* for 5 min.
4. Discard the supernatant and resuspend the cell pellet in 1 ml of DMEM.
5. Count the cells using hemocytometer and trypan blue to exclude dead cells.

6. Transfer the 1 ml of cell suspension into an eppendorf and centrifuge at 800×*g* for 5 min.
7. Discard the supernatant (see Note 9) and resuspend the cells in DMEM at a final concentration of 5×10^5 cells/5 μl (see Note 10). Keep cells on ice until injection.

3.3 Surgical Procedure

Wear lab coats, surgical gloves and masks. Clean all the working areas with ethanol wipes. Maintain aseptic conditions throughout the procedure.

1. Place the cork sheet and cover it with a sterile drape.
2. Open sterile distilled water, PBS and fill 1 ml syringe with each and label the syringes.
3. Open 2×25 g needles. These will be used to pin the mice leg and to drill a hole in the femur.
4. Open the sterile suture material, the sterile scissors, artery forceps, blades and scalpel holder from sterile pouches.
5. Weigh the mouse and intraperitoneally inject it with the ketamine–Rompun mix (see Note 6). Use 100 μl/10 g in weight of mouse.
6. Leave the mouse in the 37°C incubator covered with a sterile drape and allow 5 min for the anesthetic to act (see Note 11).
7. Pin the leg of the mice where injection is intended to the cork sheet with the 25 g needle and shave off the hairs using the electrical shaver (see Note 12) (Fig. 1a).
8. Make a small deep incision using the sterile scalpel blade above the knee joint and expose the kneecap slowly using sterile forceps by separating the tissue around without damaging any blood vessels (see Note 13) (Fig. 1b).
9. While holding the top part of the femur with sterile forceps, drill a hole gently through the groove of the kneecap in the femur about half a centimeter deep (see Note 14) (Fig. 1c).
10. Resuspend the cells gently (avoiding bubble formation) and aspirate 5 μl using the Hamilton syringe. Gently, place the needle of the Hamilton syringe in the hole and slowly inject 5 μl pushing the piston very slowly and gently (see Note 15).
11. While removing the needle, immediately close the holes with bone wax using a sterile scalpel blade.
12. Finally, wash the surgical area with sterile distilled water. Wipe the excess water using ethanol wipes and close the wound using the appropriate suture.
13. Leave the mice in 37°C incubator and monitor for recovery. Once the mice have recovered from the anesthesia, transfer them to sterile cages and monitor for infection until the date of sacrifice (5 weeks or longer).

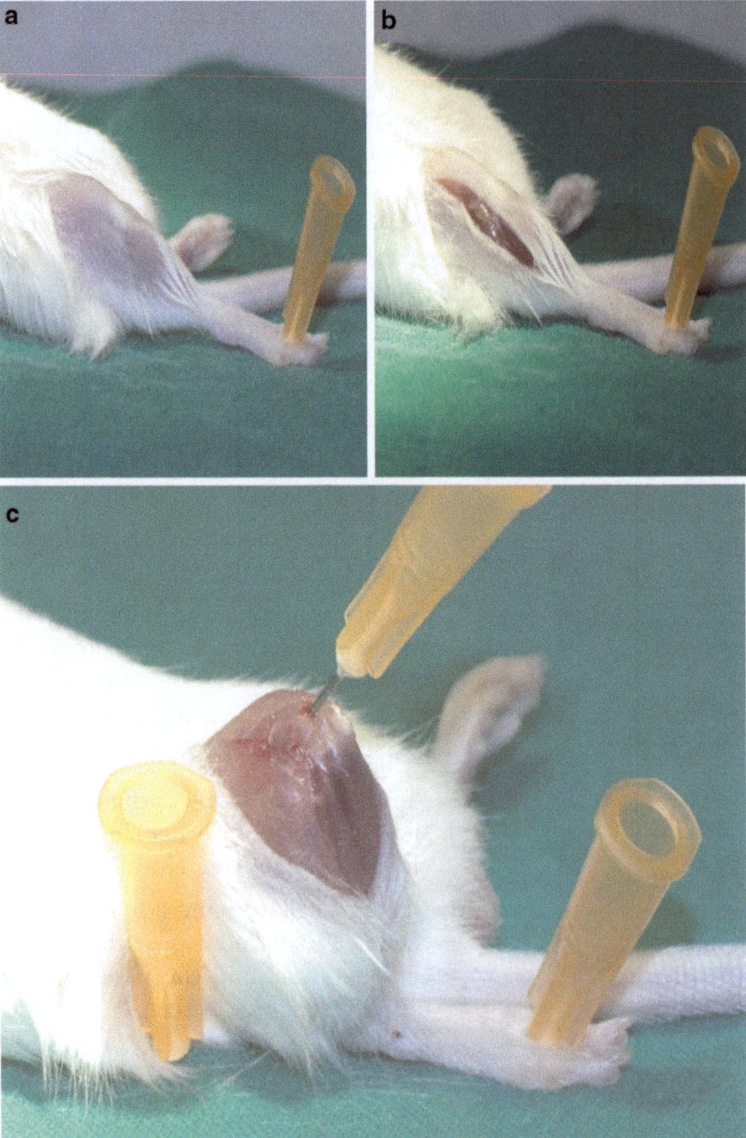

Fig. 1 Surgical procedure involving intra-femoral injections of eGFP labelled hMSC cells. (**a**) A representative image of area where the fur of the mouse has been shaved at the knee joint using an electrical shaver. (**b**) shows a 1 cm deep incision using a sterile scalpel blade aside of the knee joint and (**c**) shows the position where the hole is drilled gently through the groove of the kneecap to allow injection of hMSC cells

3.4 Immuno-Staining

To detect the levels of engraftment animals are sacrificed and bones are processed to assess the number of cells expressing eGFP by immuno-staining.

1. The femurs of mice are collected in 10% neutral buffered formalin and kept for 16 h on shaker at 4°C.

Table 1
Processing of tissue section on the Leica TP1020 carousel tissue processor

Station	Solution	Time	Vacuum
2	70% Ethanol	2 h	No
3	70% Ethanol	2 h	No
4	70% Ethanol	2 h	No
5	95% Ethanol	2 h	No
6	95% Ethanol	2 h	No
7	100% Ethanol	2 h	No
8	100% Ethanol	2 h	No
9	Xylene	2 h	No
10	Xylene	2 h	No
11	Histology wax melting point 56°C	2 h	Yes
12	Histology Wax (as above)	2 h	Yes
		Total 22 h plus 15 min drain time	

2. Following 16 h bones are decalcified in Leica decalcifier II at room temperature on a shaker for 2 h. Add 10–20 times volume of decalcifier to volume of bone.
3. On completion of decalcification, bones are transferred to labelled tissue cassettes in 70% ethanol. The tissues are immediately processed on the Leica TP1020 carousel tissue processor for paraffin embedding (Table 1).
4. After 22 h of processing, the femur is orientated longitudinally and embedded in molten wax and trimmed very slowly on the Leica RM2265 rotary microtome at 3 μm until the full head of the femur is exposed.
5. The exposed paraffin block surface is cooled for 30–60 min on an ice block stored at 4°C.
6. 25 serial sections of 3 μm thickness are cut and transferred to a 45°C distilled water bath and allowed to float for up to 30 min to avoid contraction of the bone marrow and endocortical bone.
7. The serial sections are mounted on SuperFrost Plus slides.
8. Slides are placed on a tray and dried on a hotplate at 45°C for 30 min before further drying at 37°C overnight.

9. Cooled slides are stored at 4°C before performing anti-GFP staining (no longer than 2 weeks).
10. The slides containing tissue sections are first dewaxed in xylene two times for 5 min each.
11. The tissue sections are rehydrated in 99% ethanol two times for 5 min each following which endogenous peroxidase is blocked using 3% hydrogen peroxide for 10 min (see Note 16).
12. The slides are then washed in distilled water for three times for 1 min each.
13. Wash the slides in PBS for three times for 2 min each.
14. Prepare 10% normal goat serum in PBS and incubate the tissue sections by covering the surface for 30 min at room temperature.
15. Incubate the tissue section with 200 μl primary antibody (anti-GFP, rabbit IgG fraction) overnight (see Note 17) at 4°C.
16. Wash the slides with PBS twice for 5 min each.
17. Incubate the tissue section with 200 μl secondary antibody (Goat anti-rabbit horseradish peroxidase) for 45 min at room temperature (see Note 18).
18. Wash the slides with PBS twice for 5 min each.
19. Prepare the reagents in the vector kit solution (see Note 19), add the solution (enough to cover the tissue surface) and incubate for 15 min.
20. Rinse the slides in distilled water and wash for 5 min under running tap water.
21. Counterstain nuclei in Gill's Hematoxylin for 20 s and then wash the stain with running tap water for 4 min.
22. Dehydrate the tissue sections in the order of 70, 95 and 99% industrial methylated spirit for 1 min each.
23. Finally to remove any excess trace of water the tissue sections are washed in xylene for 1 min each.
24. Clean any remaining xylene surrounding the slide using a soft tissue paper. Place a drop of DPX on a coverslip and gently invert the slides on top of it and press it gently to remove any trapped air bubble and allow the slides to dry at room temperature.
25. The eGFP stained cells will appear brown in color (see Fig. 2) and can be evaluated microscopically.

4 Notes

1. This procedure can also be carried out with freshly isolated human MSC (e.g., Lin− CD45− LNGFR+ bone marrow cells), murine MSC or HSC when numbers are small. The strain of mice used will depend on the type of cells injected.

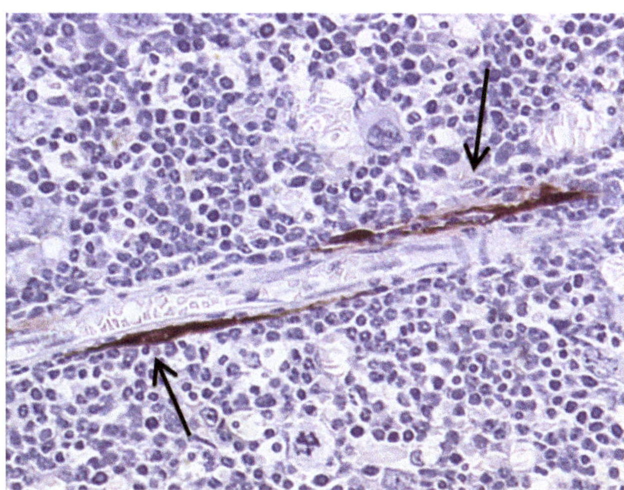

Fig. 2 eGFP+ cells in the femur of mice 5 weeks after transplantation. Representative examples of eGFP positive cells located around the vessels (resembling pericytes)

2. If an in-house system for producing lentiviral particle is not available, lentiviral vectors expression eGFP can be purchased from a number of vendors including Thermo Scientific, Sigma-Aldrich, CellBio Labs and viral particle produced according to manufacturer instructions. Lentiviral vectors are recommended over other systems due to the higher efficiency of transduction and integration of the expression vector in the genome of the cells for long term monitoring. Alternatively, other ways to detect transplanted cells can be used depending on the type of cells transplanted. For example to detect human MSC transplanted in mice an antibody to human nuclei can be used.
3. All the surgical equipments should be autoclaved at 121°C for 20 min.
4. Those needles are used to administer the anesthetic to the animals, and to clean the surgical field with PBS.
5. Syringes needs to be clean and sterilized by aspirating water first and discarding it. Repeat the operation 3 times. Then wash with ethanol 95% three times and then with PBS three times.
6. Ketaset and Rompun are anesthetics which are delivered by intraperitoneal injection. Ketaset is a regulated drug and therefore should be kept out of reach when not in use. 1 ml of Ketaset contains 100 mg/ml (equivalent to ketamine hydrochloride 115.36 mg/ml) with 0.01% benzethonium chloride as a preservative. Rompun is a concentration of 2% w/v and each ml contains xylazine hydrochloride (equivalent to 20 mg xylene) as active substance and 1.5 mg methyl-4-hydroxy-benzoate as preservative. The Ketaset and Rompun mix is prepared

at a volume of 0.5 ml of Ketaset, 0.25 ml of Rompun, and 4.3 ml of sterile distilled water.

7. Wear gloves and mask and prepare in containment hood.
8. Do not incubate the cells in trypsin/EDTA for longer time. After 2 min observe the cells under the microscope and if MSC are circular in morphology gently brisk the flask and collect the cells.
9. Ensure that when discarding the supernatant, the cell pellet is devoid of any medium. Any excess medium is removed using a sterile pipette. Avoid bubble formation while gently mixing the cell suspension in the small volume. Transfer the contents into a sterile eppendorf and store it on ice to reduce clumping of cells.
10. Prepare cells in excess of the number of mice to inject, at least 10^6 hMSC in excess to account of loss while pipetting. Moreover, if injecting cells from different donors, take into consideration that engraftment levels decrease with increasing numbers of population doublings.
11. To ensure the mouse is under the effect of the anesthetic, pinch the tip of one of the legs. If the mouse twitches, leave the mouse for few more minutes before starting the surgical procedure. Only when the mouse becomes unresponsive to the pinching, the procedure can be initiated.
12. Shave half a centimeter above and below the knee joint.
13. While making incision, if the blood vessels are cut pour sterile PBS and wipe the excessive blood using ethanol wipes and wait for a while for the blood flow to stop and then continue with the procedure.
14. Once the hole has been drilled, let pressure to ease and bone marrow to ooze out. Use sterile PBS to drain off excessive blood/marrow and use ethanol wipes to clean the blood. Always maintain aseptic conditions and keep as clean as possible to avoid infection.
15. Injecting the cells will create high pressure in the bone marrow cavity. To avoid cells leaking out, and therefore introducing greater variability in engraftment rates, cells need to be injected very slowly over 30 s. At the end of the injection leave the syringe and needle in position for 5 s before withdrawing it to stop cells exiting the bone marrow cavity.
16. Place the microscopic slides in an immuno-tray and use a Pasteur pipette to add hydrogen peroxide enough to cover the surface of the tissue section in the microscopic slide.
17. Primary antibody is prepared at a dilution of 1:600 in 5% normal goat serum. 5% normal goat serum is prepared by dissolving 1 ml of goat serum in 4 ml of PBS.

18. Secondary antibody is prepared at a dilution of 1:400 in PBS.
19. Vector kit solution is prepared for immediate use: to 5 ml of distilled water add three drops of reagent 1 (mix well), add two drops of reagent 2 (mix well), add two drops of reagent 3 (mix well), and then add two drops of hydrogen peroxide and mix well.

References

1. Rossi DJ et al (2005) Cell intrinsic alterations underlie hematopoietic stem cell aging. Proc Natl Acad Sci USA 102(26):9194–9199
2. Pittenger MF et al (1999) Multilineage potential of adult human mesenchymal stem cells. Science 284(5411):143–147
3. Friedenstein AJ, Chailakhyan RK, Gerasimov UV (1987) Bone marrow osteogenic stem cells: in vitro cultivation and transplantation in diffusion chambers. Cell Tissue Kinet 20(3):263–272
4. Baxter MA et al (2004) Study of telomere length reveals rapid aging of human marrow stromal cells following in vitro expansion. Stem Cells 22(5):675–682
5. Daga A et al (2002) Enhanced engraftment of EPO-transduced human bone marrow stromal cells transplanted in a 3D matrix in non-conditioned NOD/SCID mice. Gene Ther 9(14):915–921

Chapter 11

Tracking of Replicative Senescence in Mesenchymal Stem Cells by Colony-Forming Unit Frequency

Anne Schellenberg, Hatim Hemeda, and Wolfgang Wagner

Abstract

Long-term culture of mesenchymal stem cells (MSC) has major impact on cellular characteristics and differentiation potential. Numerous clinical trials raise high hopes in regenerative medicine and this necessitates reliable quality control of the cellular products—also with regard to replicative senescence. The maximum number of population doublings before entering the senescent state depends on the cell type, tissue of origin, culture medium as well as cell culture methods. Therefore, it would be valuable to predict the remaining proliferative potential in the course of culture expansion. Here, we describe a refined fibroblastic colony forming unit (CFU-f) assay which can be performed at any passage during culture expansion with simple cell culture techniques. This method is based on limiting dilutions in the 96-well format to determine the proportion of highly proliferative and clonogenic cells. The number of CFU-f declines rapidly during culture expansion. Especially at higher passages the CFU-f frequency correlates very well with the remaining cumulative population doublings. This approach can be used as quality measure to estimate the remaining proliferative potential of MSC in culture.

Keywords Mesenchymal stem cells, Colony forming units, Long-term culture, Senescence, Cellular aging, Limiting dilution

1 Introduction

Like all primary cells, mesenchymal stem cells (MSC) have a limited ability to proliferate *in vitro*. After a certain number of cell divisions they reach a senescent state. This phenomenon was first described over 50 years ago by Leonard Hayflick (1). Senescent cells are mitotically arrested and characterized by enlarged cell size with typical "fried-egg-morphology." More importantly cellular aging of MSC has functional consequences that impair therapeutic applications. It has been demonstrated that MSC gradually lose their adipogenic and osteogenic differentiation potential upon long-term culture (2–6). Furthermore, coculture experiments with hematopoietic stem and progenitor cells (HSC) revealed that the

Kursad Turksen (ed.), *Stem Cells and Aging: Methods and Protocols*, Methods in Molecular Biology, vol. 976,
DOI 10.1007/978-1-62703-317-6_11, © Springer Science+Business Media, LLC 2013

hematopoiesis supportive function changes over the passages: MSC from early passages maintain the primitive state of HSC for more cell divisions (7). There is also evidence that secretory and immunomodulatory functions change in aged MSC (8, 9). Therefore determining the state of cellular aging of MSC is relevant for quality control of therapeutic cell preparations (10, 11).

To date several markers are available that can detect replicative senescence of cells *in vitro*—but none of them can reliably predict the remaining proliferative potential. Progressive telomere shortening upon every cell division is considered to be one of the main triggers for replicative senescence and may serve as internal clock for the prospective replicative potential. There is evidence that quantitative analysis of telomere length correlates with the state of replicative senescence but this method does not provide reliable predictions for the remaining proliferative potential (12–14). The most widely used biomarker for senescent cells is senescence-associated beta-galactosidase (SA-β-Gal) (15). This enzyme is detectable by cytochemical staining at pH 6 due to an increase in lysosomal biogenesis which is commonly observed in senescent cells (16). However, it should be noted that β-Gal is also induced by stress during *in vitro* culture and absolute quantification of the senescent state remains difficult since it stains predominantly very large senescent cells (17).

Long-term culture of MSC is also associated with continuous gene expression changes (4, 18). Many of these are consistent in MSC isolated in different laboratories and under different culture conditions (11, 19). Furthermore, we demonstrated that long-term culture is also associated with epigenetic modifications, such as DNA-methylation at specific CpG sites (20). Senescence-associated DNA-methylation changes are enriched in developmental genes and correlate with repressive histone marks (6). We identified a subset of CpG sites which can be used to predict the number of passages, cumulative population doublings and time since culture initiation (21). This method is relatively universally applicable for different cell preparations but it may be demanding for laboratories that have no expertise in epigenetic analysis. Furthermore, the epigenetic senescence signature correlated better with the time since culture initiation than with the remaining population doublings until the cells enter replicative senescence.

Analysis of fibroblastoid colony forming unit (CFU-f) frequency provides an alternative measure for the proliferative potential during culture expansion (22, 23). The characteristic of CFU-f formation was already described in the first descriptions of MSC by Friedenstein and coworkers (24). In the classical CFU-f assays a defined cell number is seeded in a large well and the CFU-f frequency is estimated by counting discrete colonies—each of these is

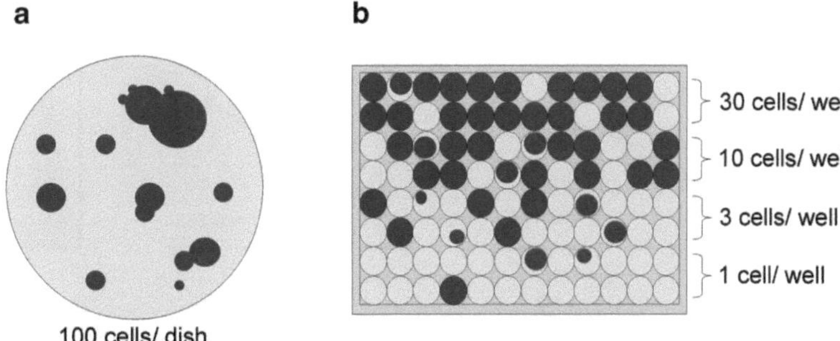

Fig. 1 Analysis of CFU-f frequency. Counting the number of colonies in the classical CFU-f assay is hampered by the need to discriminate between colonies, multi-cell derived colonies, migratory activity, and differences in proliferative activity (**a**). Alternatively, the CFU-f assay can be performed in limiting dilution in wells seeded with 1, 3, 10, and 30 cells per well. After 14 days the percentage of cells with colony formation is scored for each dilution step and the CFU-f frequency is then calculated based on Poisson statistics (**b**; see Subheading 3.3)

considered to be single cell derived. However, this approach has several disadvantages: (i) it is often difficult to distinguish between colonies, (ii) single colonies may be derived from multiple cells, (iii) cells possess migratory activity and may form daughter colonies, and (iv) slow proliferating colonies will be outgrown by faster proliferating clones and this may influence the results—especially at higher passages with increasing heterogeneity of replicative senescence (Fig. 1a) (23, 25). CFU-f frequency is usually determined during initial isolation from primary tissue but it can also be analyzed throughout culture expansion—although the entire cell population reveals plastic adherent growth only a small proportion is capable of colony formation. This proportion declines rapidly with every passage (6, 23).

In this manuscript, we describe how reliable quantification of CFU-f frequency upon passaging can be used to estimate the number of remaining cell divisions during culture expansion. The method is performed in 96-well format, where cells are seeded in limiting dilution of 30, 10, 3, and 1 cell per well. After 2 weeks the percentage of wells with CFU-f formation is determined for each dilution step using Poisson statistics (Fig. 1b). At culture initiation there is hardly any correlation between CFU-f frequency and the remaining population doublings, but after a few passages a strong positive correlation can be observed (Fig. 2a, b). This method can be easily applied in cell culture laboratories to provide an indicator for the remaining proliferative potential of MSC. For comparison we also describe the fluorescence-based assay to detect SA-βGal activity using flow cytometry (Fig. 2c).

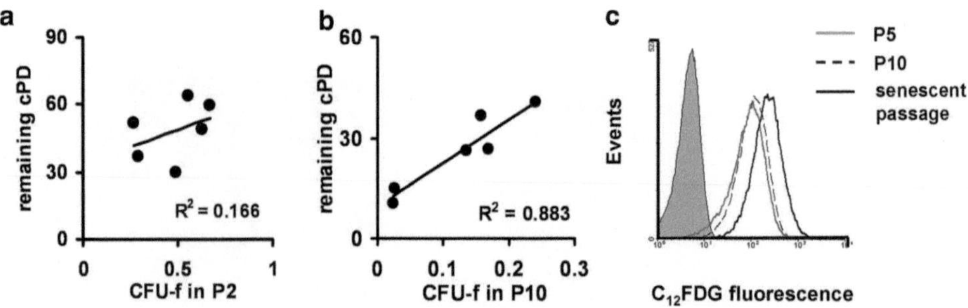

Fig. 2 Monitoring the state of senescence in MSC during long-term culture. About 20–70% of the cells at passage 2 are capable of outgrowth and colony formation (**a**). This percentage declines continuously throughout culture expansion and at passage 10 only 2–30% of the cells form colonies (**b**). CFU-f frequencies were plotted against the number of remaining population doublings until senescence. At passage 10 there is a strong positive correlation between the colony forming efficiency and the remaining proliferative potential (**b**). MSC in very late passage display higher SA-βGal activity, whereas cells of passage 5 or 10 revealed little difference (**c**)

2 Materials

2.1 Biological Material

MSC from adipose tissue (MSC-AT) were isolated from lipoaspirates from healthy adult donors after written consent according to the guidelines of the Ethics Committee of the University of Aachen (Permit number 163/07).

2.2. Generation of Human Platelet Lysate

1. Platelet units of individual healthy donors were generated by apheresis using the Trima Accel Collection System (CaridianBCT, Garching, Germany). Each thrombocyte concentrate comprises approximately $2.0–4.2 \times 10^{11}$ platelets in 200 ml plasma supplemented with Acid-Citrate-Dextrose (ACD; 1:11 v/v).
2. 0.2 μm GD/X PVDF filters (Whatman, Dassel, Germany).
3. 50 ml Falcon tube (Becton Dickinson (BD), San Jose, USA).
4. 20 ml syringe (Terumo, Eschborn, Germany).
5. Heparin-Sodium (Ratiopharm, Ulm, Germany).

2.3 Culture Expansion of MSC

1. Complete human platelet lysate (HPL) culture medium: Dulbecco's Modified Eagles Medium-Low Glucose (DMEM-LG; PAA, Pasching, Austria), 2 mM L-glutamine (Sigma-Aldrich, St Louis, MO, USA; Sigma), 100 U/ml penicillin/streptomycin (pen/strep; Gibco, Invitrogen, Carlsbad, CA), 10% pooled human platelet lysate (HPL; see Subheading 3.1), 0.61 U/ml heparin (Ratiopharm).
2. 0.25% Trypsin-EDTA solution (1×; Gibco).
3. 1× phosphate-buffered saline (PBS; PAA).
4. Tissue culture flasks 75 cm² (Nunc Thermo Fisher Scientific, Langenselbold, Germany).

5. Hemocytometer (Neubauer counting chamber, Brand, Wertheim, Germany).
6. 0.4% Trypan Blue solution (Sigma-Aldrich).

2.4 Limiting Dilution Assay

1. Crystal violet (3%) (Sigma-Aldrich) dissolved in methanol (stored at room temperature). Dilute to 0.5% working solution in Methanol prior to use (see Note 1).
2. 96-well tissue culture plate (Nunc Thermo Fisher Scientific).
3. Reagent reservoir for multi-channel pipettes (STARLAB, Ahrensburg, Germany).
4. Multi-channel pipette (STARLAB).
5. L-calc software (STEMCELL Technologies, Vancouver, Canada).

2.5 Beta Galactosidase Staining

1. 6-well plates (BD).
2. 0.1 mM Bafilomycin A1 (Sigma-Aldrich) in DMSO (stable at −20°C for several months).
3. 20 mM 5-dodecanoylaminofluorescein di-β-D-galactopyranoside (C12FDG, Invitrogen, Eugene, OR, USA) stock solution dissolved in DMSO (see Note 2).
4. 10% HPL-Medium (see Subheading 3.1).
5. 0.25% Trypsin-EDTA solution (1×; Gibco).
6. 1× phosphate-buffered saline (PBS; PAA).
7. 5 ml polystyrene test tubes suitable to use at the flow cytometer (BD).
8. FACSCanto II flow cytometer (BD) with FACSDiva acquisition software.
9. Windows Multiple Document Interface for Flow Cytometry (WinMDI 2.8) software.

3 Methods

3.1 Generation of Human Platelet Lysate

Human platelet lysates (HPL) is a highly efficient replacement for fetal calf serum (FCS). HPL has the advantage to be xenogeny-free without the risk of transmission of bovine pathogens or immunogenic reactions to FCS (26). Furthermore, especially MSC derived from adipose tissue proliferate much faster and reach more population doublings in serum supplemented with HPL as compared to FCS (27).

1. Aliquot platelet units in 50 ml Falcon tubes (see Note 3).
2. Freeze and re-thaw platelet units at −80°C and 37°C, respectively.
3. Repeat the freezing and re-thawing cycle twice.

4. Remove membrane fragments by centrifugation (2,600 × *g*, 30 min, RT).
5. Collect supernatant in 50 ml Falcon tubes. Fill in 20 ml and pass through 0.2 μm GD/X PVDF filters.
6. Collect filtrate in new 50 ml Falcon tube and store at −80°C until use (see Note 4).

3.2 Long-Term Culture of MSC

MSC were isolated from adipose tissue as described before (27). The first passage of initial colonies should be performed after about 7–10 days. It is important to always determine the number of cells harvested per passage and to define the seeding density to calculate the population doublings for each passage (PDP) and accordingly the cumulative population doublings throughout culture expansion (cPD) (10).

1. Seed cells after isolation from primary tissue in 75 cm^2 cell culture flasks in definite cell concentration (30,000 cells/cm^2 or more since MSC resemble only a very small fraction in freshly isolated cells).
2. Culture cells until they reach about 80% confluence (see Note 5). Remove culture medium and rinse cells with 1× PBS.
3. Remove 1× PBS, add 1 ml 0.25% trypsin-EDTA solution, and incubate for 1–3 min at 37°C (see Note 6).
4. Ensure cell detachment using a microscope and stop trypsinization by adding 6 ml culture medium.
5. Collect cell suspension by centrifugation (350 × *g*, 7 min, RT).
6. Discard the supernatant, resuspend pellet in HPL-culture medium. Count cells and determine viability by Trypan blue dye exclusion using a Neubauer counting chamber.
7. Cell population doublings per passage (PDP) and cumulative population doublings (cPD) are calculated from the first passage onward by the following formula: $\mathrm{PDP}_i = \log_2 (h_i / s_i)$; $\mathrm{cPD} = \mathrm{sum}_{i=1\ldots n} \mathrm{PDP}_i$, where n is the total number of passages, s_i is the number of cells seeded at passage i, and h_i is the number of cells harvested in passage i (Fig. 3) (see Note 7).
8. Reseed cells in 75 cm^2 cell culture flask at a density of 10,000 cells/cm^2 and proceed with Subheading 3.2, step 2.

Upon passaging only few cells are capable of new colony formation. To accommodate the fact that the progeny of each passage is based on a decreasing number of highly proliferative cells, it is possible to alternatively calculate the number of population doublings for each passage divided by the corresponding CFU-f frequency. This method of CFU-f-adjusted growth curves results

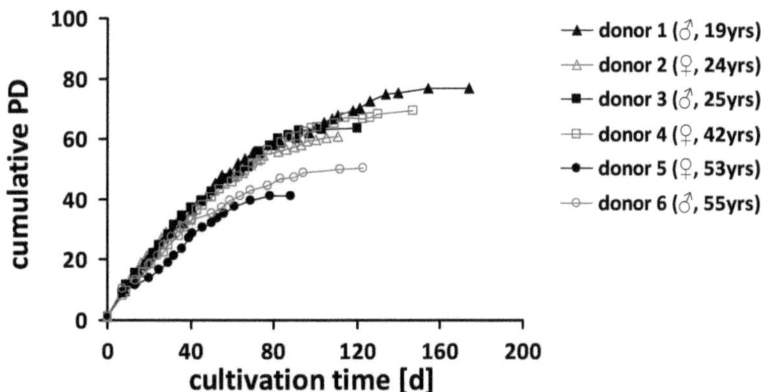

Fig. 3 Long-term growth curves of MSC. Adipose tissue-derived mesenchymal stem cells of six donors were culture expanded until they stopped proliferation and reached a senescent state. Every cell passage is indicated by a *point* and the number of cumulative population doublings (cPD) was calculated based on the ratio of cells seeded versus cells harvested per passage. To account for the fact, that the progeny of each passage is based on a decreasing percentage of highly proliferative cells it would alternatively be possible to calculate CFU-f-adopted growth curves (see Subheading 3.2)

in much higher numbers of cumulative population doublings: $PDP^{CFU\text{-}f} = \left(h_i / \left(s_i \times CFU\text{-}f_i\right)\right)$ and $cPD^{CFU\text{-}f} = sum_{i=1...n} PDP^{CFU\text{-}f}_i$ (6). This however requires CFU-f analysis at each passage.

3.3 Analysis of CFU-f

3.3.1 Analysis of Initial CFU-f

Adipose tissue-derived primary cell cultures comprise subpopulations of adherent and non-adherent cells. Taken into account that only a small proportion of cells will attach to the tissue culture surface, determination of initial CFU-f can be used to calculate population doublings prior to the first passaging. Therefore, PDP at the initial passage is calculated based on the ratio of CFU-f seeded *versus* cells harvested.

1. Count the number of isolated cells from primary tissue and determine viability by Trypan blue dye exclusion using a Neubauer chamber.
2. Dilute cells in culture medium (see Note 8) and seed cells in a 96-well plate using following limiting dilutions: 100; 300; 1,000; and 3,000 cells per well (24 replica for each condition) (see Notes 9 and 10) (23).
3. Cultivate cells at 37°C in a humidified atmosphere with 5% CO_2 with media exchange twice per week (see Note 11).
4. After 14 days, wash cells with 1× PBS and stain with 0.1% crystal violet solution.
5. Incubate cells for 10 min at room temperature (see Notes 12 and 13).

6. Wash cells at least three times with aq. dest. and score wells of each dilution step for colony-formation (Fig. 2) (see Note 14).
7. Colony formation is defined as a visible crystal violet staining that can be detected by naked eye (see Note 15).

Calculate CFU-f frequency for example by using L-Calc Software (STEMCELL Technologies): Enter values for dose (number of cells seeded per well), responses (number of wells with colony-formation per dilution step), and replicates tested (total number of wells per dilution step). The calculations realized with L-Calc Software are based on Poisson statistics. We assume that the number of CFUs follows a Poisson distribution (28–31). Denote by λ the average number of CFUs per well. The probability $p_\lambda(i)$ for exactly i CFUs (see Note 16) in a well is then given by:

$$p_\lambda(i) = \frac{e^{-\lambda}\lambda^i}{i!}$$

3.3.2 Analysis of CFU-f in Long-Term Culture

CFU-f assay applied in subsequent passages can be used as an indicator for the replicative potential (see Note 17). Here, the whole starting population reveals plastic adherent growth and therefore the percentage of CFU-f is much higher than for primary tissue. It might be unexpected that despite plastic adherent growth only a fraction of the cells is capable of new CFU-f formation. Throughout culture expansion, correlation between CFU-f and remaining population doublings increases (Fig. 2a, b).

1. Cultivate cells according to the long-term culture protocol (see Subheading 3.2).
2. Upon passaging seed cells in limiting dilutions in 96-well plates (1, 3, 10, and 30 cells per well; 24 replica for each condition) (see Note 9).
3. Carry out steps 3–6 for CFU-f analysis as described in Subheading 3.3.1.

3.3.3 Beta Galactosidase Staining

In order to detect SA-βGal activity, the lysosomal pH value needs to be raised to approximately 6 using lysosomal inhibitors such as Bafilomycin A1 (32). Thereafter the cells are incubated with C_{12}FDG that becomes fluorescent upon hydrolysis by β-galactosidase as described previously (33).

1. Seed 10,000 cells/cm2 in a 6-well plate and cultivate till 70–80% confluence (see Notes 18 and 19).
2. Remove cell culture medium and incubate cells at 37°C in a humidified atmosphere with 5% CO2 with 100 nM Bafilomycin A1 in fresh cell culture medium (2 ml per well).
3. After 1 h incubation, add 33 μl of 2 mM C_{12}FDG working solution to the same cell culture medium containing 100 nM

Bafilomycin A1 (final concentration of C_{12}FDG is ~33 μM) (see Note 20).

4. Continue the incubation at 37°C in a humidified atmosphere with 5% CO2 for 1–2 h.
5. Remove the solution and wash the cells twice with 2 ml 1 × PBS at room temperature.
6. Harvest the cells with 0.25% trypsin-EDTA (see Subheading 3.2, steps 3 and 4) and centrifuge at 350 × *g* for 7 min.
7. Resuspend the cell pellet in ice-cold PBS at a concentration of ~1 × 106 cells/ml. Approximately 0.4 ml cell suspension are required for analysis.
8. Measure the fluorescence intensity of the cells immediately in a flow cytometer. C_{12}FDG is measured on the FL-1 detector (494–518 nm wavelength) (see Note 21).
9. Analyze acquired data using WinMDI 2.8 software.
10. Process data using a histogram plot as shown in Fig. 2c in order to estimate the relative SA-βGal activity.

4 Notes

1. To reduce impurity, filter crystal violet stock solution trough 5 μm Whatman filter paper (Whatman Biometra, Göttingen, Germany).
2. Dilute 20 mM C_{12}FDG stock solution 1:10 with fresh cell culture medium directly before adding to the cells.
3. The volume will increase during the freezing procedure—do not fill the tubes with more than 45 ml.
4. Avoid multiple thawing and refreezing of HPL-aliquots to prevent loss of activity.
5. 80% confluence should not be exceeded during culture expansion to prevent contact inhibition and increased cellular stress through confluent cell growth. It has to be noted that especially in the first isolation step from primary tissue the number of colonies may be very low and then the cells need to be passaged at lower confluency level to prevent formation of few very large colonies (27).
6. Observe trypsinization under a microscope, since HPL-cultured MSC are more sensitive to trypsin treatment compared to FCS-cultured MSC.
7. It is important to determine PDP for each passage throughout culture expansion. Otherwise the cPD cannot be determined retrospectively.
8. Previous studies demonstrated that MSC culture expansion with HPL or FCS as serum supplement revealed significant

differences in regard to proliferation rate, cell size, and maximum number of PDs (27). Nevertheless, the limiting dilution assay described in this chapter is also applicable to MSC cultures supplemented with FCS.

9. It is important that cells are well dissociated to ensure seeding of single cells in all 96 wells.

10. This seeding density refers to adipose-tissue derived primary cultures. Analysis of initial CFU-f frequency of bone marrow derived MSC might require higher seeding densities (e.g., 1,000; 3,000; 10,000; 33,000; 100,000; and 333,000 cells per well; 34) due to the lower frequency of CFU-f in bone marrow.

11. Ensure that during the 14 days of incubation a constant frequency of medium exchange (twice per week) is maintained for each sample to avoid inadvertent impact on the proliferation rate.

12. Two weeks of culture expansion might be too long for the conventional CFU-f assay without limiting dilution—faster proliferating colonies would become extremely large by this time. This is an advantage of the limiting dilution approach that can take also slow growing colonies into account.

13. Alternative to crystal violet staining, MTT (3-(4,5-Dimethylthiazol-2-yl)-2,5-diphenyltetrazolium bromide) staining can be performed to visualize colony formation.

14. For high-throughput analysis, crystal violet intensity per well can be measured with a plate reader. Read absorbance of crystal violet at 590 nm.

15. In the present study, wells that showed a visible crystal violet staining upon analysis with the naked eye, were considered as wells with colony formation. It is also conceivable to set a different cutoff, e.g., score all wells with at least 50% confluent cell growth (to reach >50% confluent growth per well the cell had to be capable of at least 14 cell divisions to give rise to >15,000 cells).

16. It has to be noted that some of these colonies are not clonally derived. For calculating the frequency of monoclonally derived cells see ref. 23.

17. Additionally, monitoring of adipogenic and osteogenic differentiation potential during long-term culture can be performed with the described limiting dilution assay. Therefore, culture medium is exchanged to adipogenic or osteogenic differentiation medium after 14 days. Osteogenic differentiation can be analyzed after 3 weeks by Alizarin Red staining and quantified with a plate reader (405 nm, Tecan Infinite M200), whereas the percentage of adipogenic differentiated

cells can be determined by staining of fat droplets with BODIBY (4,4-difluoro-1,2,5,7,8-pentamethyl-4-bora-3a,4a-diaza-s-indacene) and counterstaining with DAPI (4′,6-Diamidin-2-phenylindol) as described before (23, 27).

18. Ensure subconfluence of cell culture when carrying out SA-β-Gal assay since confluent cell growth is reported to induce β-Gal activity.
19. It is recommended to include positive and negative controls for the SA-β-Gal assay. Normal MSC treated with oxidative stress might serve as positive control (35), whereas early-passage MSC represent a negative control for SA-β-Gal activity.
20. Caution: Bafilomycin is an irritant. When working with Bafilomycin use a fume hood and wear gloves and goggles.
21. If there is no green fluorescence signal detectable by flow cytometry check the experimental set up by using a fluorescence microscope.

Acknowledgement

This work was supported by the excellence initiative of the German federal and state governments within the START-Program of the Faculty of Medicine, RWTH Aachen, by the Stem Cell Network North Rhine Westphalia, and by the Else-Kröner Fresenius Stiftung.

References

1. Hayflick L, Moorhead PS (1961) The serial cultivation of human diploid cell strains. Exp Cell Res 25:585–621
2. Bonab MM, Alimoghaddam K, Talebian F, Ghaffari SH, Ghavamzadeh A, Nikbin B (2006) Aging of mesenchymal stem cell in vitro. BMC Cell Biol 7:14
3. Noer A, Boquest AC, Collas P (2007) Dynamics of adipogenic promoter DNA methylation during clonal culture of human adipose stem cells to senescence. BMC Cell Biol 8:18
4. Wagner W, Horn P, Castoldi M et al (2008) Replicative senescence of mesenchymal stem cells—a continuous and organized process. PLoS One 5:e2213
5. Banfi A, Muraglia A, Dozin B, Mastrogiacomo M, Cancedda R, Quarto R (2000) Proliferation kinetics and differentiation potential of ex vivo expanded human bone marrow stromal cells: implications for their use in cell therapy. Exp Hematol 28:707–715
6. Schellenberg A, Lin Q, Schueler H et al (2011) Replicative senescence of mesenchymal stem cells causes DNA-methylation changes which correlate with repressive histone marks. Aging (Albany NY) 3:873–888
7. Walenda T, Bork S, Horn P et al (2010) Co-culture with mesenchymal stromal cells increases proliferation and maintenance of hematopoietic progenitor cells. J Cell Mol Med 14:337–350
8. Campioni D, Rizzo R, Stignani M et al (2009) A decreased positivity for CD90 on human mesenchymal stromal cells (MSCs) is associated with a loss of immunosuppressive activity by MSCs. Cytometry B Clin Cytom 76: 225–230
9. Liang H, Hou H, Yi W et al (2011) Increased expression of pigment epithelium-derived factor in aged mesenchymal stem cells impairs their therapeutic efficacy for attenuating myocardial infarction injury. [Epub ahead of print]
10. Wagner W, Ho AD, Zenke M (2010) Different facets of aging in human mesenchymal stem cells. Tissue Eng Part B Rev 16:445–453
11. Wagner W, Bork S, Lepperdinger G et al (2010) How to track cellular aging of mesenchymal stromal cells. Aging (Albany NY) 2:224–230

12. Allsopp RC, Vaziri H, Patterson C et al (1992) Telomere length predicts replicative capacity of human fibroblasts. Proc Natl Acad Sci USA 89:10114–10118
13. Baxter MA, Wynn RF, Jowitt SN, Wraith JE, Fairbairn LJ, Bellantuono I (2004) Study of telomere length reveals rapid aging of human marrow stromal cells following in vitro expansion. Stem Cells 22:675–682
14. Fehrer C, Voglauer R, Wieser M et al (2006) Techniques in gerontology: cell lines as standards for telomere length and telomerase activity assessment. Exp Gerontol 41:648–651
15. Dimri GP, Lee X, Basile G et al (1995) A biomarker that identifies senescent human cells in culture and in aging skin in vivo. Proc Natl Acad Sci USA 92:9363–9367
16. Kurz DJ, Decary S, Hong Y, Erusalimsky JD (2000) Senescence-associated (beta)-galactosidase reflects an increase in lysosomal mass during replicative ageing of human endothelial cells. J Cell Sci 113(Pt 20):3613–3622
17. Zhou S, Greenberger JS, Epperly MW et al (2008) Age-related intrinsic changes in human bone-marrow-derived mesenchymal stem cells and their differentiation to osteoblasts. Aging Cell 7:335–343
18. Izadpanah R, Kaushal D, Kriedt C et al (2008) Long-term in vitro expansion alters the biology of adult mesenchymal stem cells. Cancer Res 68:4229–4238
19. Schallmoser K, Bartmann C, Rohde E et al (2010) Replicative senescence-associated gene expression changes in mesenchymal stromal cells are similar under different culture conditions. Haematologica 95:867–874
20. Bork S, Pfister S, Witt H et al (2010) DNA methylation pattern changes upon long-term culture and aging of human mesenchymal stromal cells. Aging Cell 9:54–63
21. Koch CM, Joussen S, Schellenberg A, Lin Q, Zenke M, Wagner W (2012) Monitoring of Cellular Senescence by DNA-Methylation at Specific CpG sites. Aging Cell 11:366–369
22. DiGirolamo CM, Stokes D, Colter D, Phinney DG, Class R, Prockop DJ (1999) Propagation and senescence of human marrow stromal cells in culture: a simple colony-forming assay identifies samples with the greatest potential to propagate and differentiate. Br J Haematol 107:275–281
23. Schellenberg A, Stiehl T, Horn P et al (2012) Population dynamics of mesenchymal stromal cells during culture expansion. Cytotherapy 14(4):401–411
24. Friedenstein AJ, Deriglasova UF, Kulagina NN et al (1974) Precursors for fibroblasts in different populations of hematopoietic cells as detected by the in vitro colony assay method. Exp Hematol 2:83–92
25. Wagner W (2010) Senescence is heterogeneous in mesenchymal stromal cells—kaleidoscopes for cellular aging. Cell Cycle 9: 2923–2924
26. Schallmoser K, Rohde E, Bartmann C, Obenauf AC, Reinisch A, Strunk D (2009) Platelet-derived growth factors for GMP-compliant propagation of mesenchymal stromal cells. Biomed Mater Eng 19:271–276
27. Cholewa D, Stiehl T, Schellenberg A et al (2011) Expansion of adipose mesenchymal stromal cells is affected by human platelet lysate and plating density. Cell Transplant 20(9):1409–1422
28. Staszewski R (1990) Murphy's law of limiting dilution cloning revisited. Stat Med 9:1541
29. Underwood PA, Bean PA (1988) Hazards of the limiting-dilution method of cloning hybridomas. J Immunol Methods 107: 119–128
30. Chamberlain JR, Schwarze U, Wang PR et al (2004) Gene targeting in stem cells from individuals with osteogenesis imperfecta. Science 303:1198–1201
31. Lietzke R, Unsicker K (1985) A statistical approach to determine monoclonality after limiting cell plating of a hybridoma clone. J Immunol Methods 76:223–228
32. Yoshimori T, Yamamoto A, Moriyama Y, Futai M, Tashiro Y (1991) Bafilomycin A1, a specific inhibitor of vacuolar-type H(+)-ATPase, inhibits acidification and protein degradation in lysosomes of cultured cells. J Biol Chem 266:17707–17712
33. Debacq-Chainiaux F, Erusalimsky JD, Campisi J, Toussaint O (2009) Protocols to detect senescence-associated beta-galactosidase (SA-betagal) activity, a biomarker of senescent cells in culture and in vivo. Nat Protoc 4:1798–1806
34. Walenda G, Hemeda H, Schneider RK, Merkel R, Hoffmann B, Wagner W (2012) Human platelet lysate gel provides a novel three dimensional-matrix for enhanced culture expansion of mesenchymal stromal cells. Tissue Eng Part C Methods 18:924–934
35. Brandl A, Meyer M, Bechmann V, Nerlich M, Angele P (2011) Oxidative stress induces senescence in human mesenchymal stem cells. Exp Cell Res 317:1541–1547

Chapter 12

Chromatin Immunoprecipitation Coupled by Quantitative Real-Time PCR as a Tool for Analyzing Epigenetic Regulation of Stem Cell Aging

Seunghee Lee*, Ji-Won Jung*, and Kyung-Sun Kang

Abstract

Epigenetic regulation is one of the major players in gene expression control. Histone modifications including acetylation or methylation are epigenetic mechanisms known to regulate senescence associated genes expression during adult stem cell aging. Chromatin immunoprecipitation (ChIP) assay is widely used to investigate histone modification patterns of genes of interest. For ChIP analysis, there are several conditions to be optimized empirically. In this article, not only a method of ChIP coupled by real-time quantitative PCR but also the detailed conditions is provided based on our previously published studies.

Keywords Adult stem cell, Cellular senescence, Epigenetic, Histone modification, Chromatin immunoprecipitation, Real-time qPCR

1 Introduction

In adult stem cell aging, epigenetic regulation of gene expression is an important mechanism which controls key players of cellular senescence (1, 2). Histone modifications by acetylation or methylation are two of the main epigenetic regulations mechanisms. The eukaryotic DNA is wrapped around core histone octamer complexes, composing one H3-H4 tetramer and two H2A-H2B (3, 4). According to the modification patterns of histones, chromatin structures change and affect transcriptions of the genes wrapped around them. N termini of histones lose their affinity to bind to DNA by neutralization of positively charged lysine residues mediated by acetylation. Then the N termini are allowed to displace from the nucleosome allowing the nucleosome to unfold and increasing accessibility for transcription factors (5). In case of methylation, the transcriptional activity is determined according to the location of lysine residue which is

*Seunghee Lee and Ji-Won Jung contributed equally to this work.

Kursad Turksen (ed.), *Stem Cells and Aging: Methods and Protocols*, Methods in Molecular Biology, vol. 976, DOI 10.1007/978-1-62703-317-6_12, © Springer Science+Business Media, LLC 2013

methylated. Histone H3 lysine 4 trimethylation (H3K4me3), has been shown as a major conserved mark of transcriptionally active chromatin (6). Histone H3 lysine 27 trimethylation (H3K27me3) is a marker associated with transcriptional silencing, in particular, of developmental transcription factors (7).

During replicative senescence of mesenchymal stem cells, various promoter regions are changed epigenetically. The promoter region of p16INK4A, a well-known senescence marker, is changed into transcription active forms resulting in the increase of p16INK4A expression (8). Not only the protein coding genomic DNA but also the genomic DNA regions of noncoding small RNAs are regulated by epigenetic changes (9). Chromatin immunoprecipitation (ChIP) assay is a useful method in investigating these histone modifications. ChIP is an experimental technique used to investigate the interaction between protein and DNA in the cell. Since the first try to investigate the interactions between proteins and nucleic acid (10, 11), the technique of "chromatin immunoprecipitation" has been developed extensively and refined thereafter. To investigate protein–DNA interactions, the protein–DNA complex should be properly cross-linked using cross-linking reagent such as formaldehyde. The DNA is then sheared to smaller pieces (200~1,000 bp) by sonication or nuclease digestion. After that, the complexes of protein and sheared DNA are immunoprecipitated by a highly specific Ab against the protein. A certain volume ratio of the sheared DNA before immunoprecipitation is kept as an input control sample. The protein–DNA complexes from input control and ChIP samples are reverse cross-linked using high concentration of NaCl. After purification the DNA can be analyzed using a number of techniques, such as quantitative PCR, sequencing or microarray for presence of nucleotide sequence of interest (12).

Although ChIP is widely used technique to investigate epigenetic histone marks in genomic DNA, there are several conditions to be optimized by an experimenter empirically. It is intricate and time consuming to initially setup ChIP procedures. In this article, we describe a method of ChIP for investigation of histone modifications of p16INK4A, a gene related to adult stem cell aging, coupled by real-time qPCR. The protocol described here is based on a manufacturer's instruction for a commercially available ChIP kit (Millipore, USA) with a number of modifications to fit the aims of our study. We also describe detailed conditions for ChIP optimization based on our previous publications.

2 Materials

2.1 Chromatin Immunoprecipitation Components

1. Salmon Sperm DNA/Protein A or G Agarose: Add 1 mg BSA and 400 μg salmon sperm DNA into 1 ml Protein A or G Agarose beads (Protein A Agarose: Catalog# 16-125, Protein G Agarose: Catalog# 16-266, Millipore) (see Note 1).

2. ChIP Dilution Buffer: 0.01% SDS, 1.1% Triton X-100, 1.2 mM EDTA, 16.7 mM Tris–HCl, 167 mM NaCl, pH 8.1. When you prepare 100 ml Buffer, dissolve 1 mg SDS, 1.1 ml Triton X-100, 44.67 mg EDTA, 263.18 mg Tris–HCl, and 977 mg NaCl in 100 ml DW. After dissolving the compounds thoroughly, filtrate and adjust acidity to pH 8.1 (see Note 2).
3. Low Salt Immune Complex Wash Buffer: 0.1% SDS, 1% Triton X-100, 2 mM EDTA, 20 mM Tris–HCl, 150 mM NaCl, pH 8.1. When you prepare 100 ml Buffer, dissolve 0.1 mg SDS, 1 ml Triton X-100, 74.45 mg EDTA, 315.18 mg Tris–HCl, and 877.5 mg NaCl in 100 ml DW. After dissolving the compounds thoroughly, filtrate and adjust acidity to pH 8.1 (see Note 3).
4. High Salt Immune Complex Wash Buffer: 0.1% SDS, 1% Triton X-100, 2 mM EDTA, 20 mM Tris–HCl, 500 mM NaCl, pH 8.1. When you prepare 100 ml Buffer, dissolve 0.1 mg SDS, 1 ml Triton X-100, 74.45 mg EDTA, 315.18 mg Tris–HCl, and 2.93 g NaCl in 100 ml DW. After dissolving the compounds thoroughly, filtrate and adjust acidity to pH 8.1 (see Note 3).
5. LiCl Immune Complex Wash Buffer: 0.25 M LiCl, 1% IGEPAL-CA630, 1% deoxycholic acid (sodium salt), 1 mM EDTA, 10 mM Tris–HCl, pH 8.1. When you prepare 100 ml Buffer, dissolve 1.06 g LiCl, 1 ml IGEPAL-CA630, 1 mg deoxycholic acid sodium salt, 37.23 mg EDTA, and 157.59 mg Tris–HCl in 100 ml DW. After dissolving the compounds thoroughly, filtrate and adjust acidity to pH 8.1 (see Note 3).
6. TE Buffer: 10 mM Tris–HCl, 1 mM EDTA, pH 8.0. When you prepare 100 ml Buffer, dissolve 157.59 mg Tris–HCl and 37.23 mg EDTA in 100 ml DW. After dissolving the compounds thoroughly, filtrate and adjust acidity to pH 8.0 (see Note 4).
7. 0.5 M EDTA, pH 8.0.
8. 5 M NaCl.
9. 1 M Tris–HCl, pH 6.5.
10. SDS Lysis Buffer: 1% SDS, 10 mM EDTA, 50 mM Tris, pH 8.1.
11. Antibody for chromatin immunoprecipitation.
12. 37% Formaldehyde.
13. PBS.
14. Elution buffer: 1% SDS, 0.1 M $NaHCO_3$. When you prepare 20 ml buffer, dissolve 200 mg SDS and 168 mg $NaHCO_3$ in 20 ml DW and filtrate it. Make this solution freshly on 2nd working day.
15. Molecular Biology grade Proteinase K.
16. Glycogen or tRNA.

2.2 DNA Extraction

1. 50% Phenol/50% Chloroform (containing 1% isoamyl alcohol).
2. 100% Isopropanol.
3. Sodium acetate.
4. 70% Ethanol.

3 Methods

3.1 Optimization of DNA Shearing

According to the cell type, cell number, volume of cell lysates, DNA shearing condition should be independently optimized. Establishment of optimal conditions is important to obtain reliable result. The ideal size of DNA is range from 200 to 1,000 bp in length for PCR analysis.

1. Prepare cell number five times larger than you want in order to set up the condition. If you want to set up 1×10^6 cells condition, prepare 5×10^6 cells. To estimate cell number, one extra dish should be prepared (see Note 5).
2. Cross-link histones to DNA by adding formaldehyde directly to culture medium to a final concentration of 1% and incubate for 10 min at 37°C. (For example, add 270 μl 37% formaldehyde into 10 ml of growth medium on plate) (see Note 6).
3. To stop the cross-linking reaction, add 2.5 M Glycine as 1/20 volume of medium.
4. Aspirate medium, removing as much medium as possible. Wash cells twice using ice-cold PBS containing protease inhibitors (1 mM phenylmethylsulfonyl fluoride (PMSF), 1 μg/ml aprotinin, and 1 μg/ml pepstatin A) (see Note 7).
5. Scrape cells into conical tube.
6. Pellet cells for 3 min at 16,000 × *g* at 4°C. Warm SDS Lysis Buffer to room temperature to dissolve precipitated SDS and add protease inhibitors cocktail (1 mM PMSF, 1 μg/ml aprotinin, and 1 μg/ml pepstatin A).
7. Resuspend cell pellet in SDS Lysis Buffer and incubate for 10 min on ice. If the cell number is 1×10^6, resuspend cell pellet in 200 ml SDS lysis Buffer (see Note 8).
8. Sonicate lysate to shear DNA to lengths between 200 and 1,000 bp being sure to keep samples ice cold. Once sonication conditions have been optimized, keep cell number consistent for subsequent experiments (see Note 9).
9. Add 8 μl 5 M NaCl and reverse cross-links at 65°C for overnight.
10. Recover DNA by phenol/chloroform extraction as following Subheading 3.3 and run 10 μl of each sample in an agarose gel to visualize shearing efficiency (see Note 10).

3.2 Chromatin Immunoprecipitation Protocol

1. Calculate the cell number as follows and prepare cells for ChIP analysis.

$$[1\times 10^{6}\,\text{cells} / \text{sample}]\times[\text{sample number}]$$

$$^{*}\text{sample number : target protein number} + 1(\text{for negative control})$$

   If you want to see three histone marks (e.g., acetyl histone H4, histone H3K4Me3, histone H3K27Me3), prepare 4×10^{6} cells.

2. Process the samples by completing steps 2–7 in the Subheading 3.1 and continue to the steps described hereafter.
3. Sonicate lysates to shear DNA to lengths between 200 and 1,000 bp using optimized sonication conditions in the Subheading 3.1 (see Note 5).
4. Centrifuge samples (from Subheading 3.1, step 8) for 10 min at $16,000\times g$ at 4°C, and transfer the supernatant to a new 2 ml-microcentrifuge tube. Discard pellet.
5. Dilute the sonicated cell supernatant tenfold in ChIP Dilution Buffer containing protease inhibitors cocktail as above. This is done by adding 1.8 ml ChIP Dilution Buffer to the 200 μl sonicated cell supernatant for a final volume of 2 ml.
6. To proceed PCR analysis, a portion of the diluted cell supernatant 1% (~20 μl) should be kept to quantify the amount of DNA present in different samples at the PCR protocol. These samples are considered to be your input/starting material and needs to have the histone–DNA cross-links reversed by adding 1 μl of 5 M NaCl and heating at 65°C for 4 h (see step 14). Keep these input samples at −20°C until step 14.
7. To reduce nonspecific background, pre clear the 1.8 ml diluted cell supernatant with 60 μl of Salmon Sperm DNA/Protein A Agarose-50% Slurry for 30 min at 4°C with agitation (see Note 1).
8. Pellet agarose by brief centrifugation. Three samples are for specific immunoprecipitation and one sample is for negative control (see Note 11).
9. Add an antibody for immunoprecipitation (the amount of antibody should follow the manufacturer's recommendation or decided individually by researchers) to the 2 ml supernatant fraction and incubate overnight at 4°C with rotation. For a negative/background control, prepare a sample to use as a no-antibody immunoprecipitation control.
10. Add 45 μl of Salmon Sperm DNA/Protein A Agarose Slurry for 1 h at 4°C with rotation to collect the antibody/histone complex.

11. Pellet agarose by gentle centrifugation (100 × *g* at 4°C, ~1 min). Carefully remove the supernatant that contains unbound, nonspecific DNA. Wash the protein A agarose/antibody/histone complex for 3–5 min on a rotating platform with 1 ml of each of the buffers listed in the order as given below:
    (a) Low Salt Immune Complex Wash Buffer, one wash
    (b) High Salt Immune Complex Wash Buffer, one wash
    (c) LiCl Immune Complex Wash Buffer, one wash
    (d) TE Buffer, two washes
12. Freshly prepare elution buffer (1% SDS, 0.1 M $NaHCO_3$).
13. Elute the histone complex from the antibody by adding 250 μl elution buffer to the pelleted protein A agarose/antibody/histone complex from step 11d above. Vortex briefly to mix and incubate at room temperature for 15 min with rotation. Spin down agarose and carefully transfer the supernatant fraction (eluate) to another tube and repeat elution. Combine eluates (total volume ~ 500 μl).
14. Add 20 μl 5 M NaCl to the combined eluates (500 μl) and reverse histone–DNA cross-links by heating at 65°C for 4 h. Include the input/starting material (the sample saved from Subheading 3.2, step 6), and add 1 μl of 5 M NaCl per 20 μl sample and reverse crosslink with other samples.

3.3 Recovering DNA

1. Add 10 μl of 0.5 M EDTA, 20 μl 1 M Tris–HCl, pH 6.5, and 2 μl of 10 mg/ml Proteinase K to the combined eluates and incubate for 1 h at 45°C.
2. Mix same volume of PCI mixture (500 μl) and eluates (500 μl) with vortexing or tapping. Centrifuge samples for 15 min at 16,000 × *g* at 4°C (see Notes 12 and 13).
3. Transfer the aqueous phase to a new tube and add 0.1 volume of 3 M sodium acetate and 1 volume of isopropyl alcohol. Shake the mixture and incubate for 1 h at −20°C.
4. Centrifuge for 15 min at 16,000 × *g* at 4°C and remove the upper layer. Wash the pellet with 1 ml 75% EtOH.
5. Spin-down pellet for 10 min at 16,000 × *g* and air dry to remove remained EtOH.
6. Dilute pellet with 20 μl DEPC treated or distilled water.

3.4 PCR Protocol to Amplify DNA That Is Bound to the Immunoprecipitated Histone

1. Design primer pairs to produce an amplicon ranges 50~150 bp in length (see Note 14).
2. Set up real-time PCRs in triplicate with a diluted input and immunoprecipitated DNA, as shown in the following table (see Notes 15 and 16).

Component	Amount per reaction (μl)	Final concentration
Primer mix (stock 5 μM each)	1	0.4 μM each
Diluted DNA (input or sample)	1	20× diluted
Nuclease free water	8	
SYBR Green mix	10	

3. Perform real-time PCR using cycling conditions as shown in the following table. Make sure that the program calculates the dissociation curve after the completion of PCR. C_t values for the technical repeats should be evaluated for quality of PCR steps.

	Stage1	Stage2	Stage3	
Temperature	50.0°C	95.0°C	95.0°C	60.0°C
Time	2 min	10 min	15 s	1 min
Cycles	1	1	40	

4. Calculate the DC_t value (normalized to the input samples) for each sample: DC_t (C_t (sample) − C_t (input)). Next, calculate the DDC_t (DC_t (experimental sample) − DC_t (negative control)). Calculate the fold difference between experimental sample using $2^{(-\mathrm{DD}C_t)}$.

4 Notes

1. Experimenter should select Protein A or G agarose bead according to the antibody isotype as following Table 1.
2. The volume of ChIP dilution buffer for experiment: (2 ml/reaction) × (Number of reaction).

   With an exception for Subheading 2.1, item 14, buffers are stable in 4°C for 1 year. So, experimenter can make enough solution and keep in 4°C for 1 year.
3. The volume of Low, High Salt or LiCl Immune Complex Wash Buffer for experiment: (1 ml/reaction) × (Number of reaction)
4. The volume of TE Buffer for experiment: (1 ml/reaction) × (Number of reaction) × 2
5. 1×10^6 cells are usually used for one chromatin immunoprecipitation reaction. If you set up the sonication condition of this cell number, you have to apply this condition to only 1×10^6 cells. So, if the target proteins are increased, you have to divide cells into several tubes to maintain 1×10^6 cells per

Table 1
Binding capability of Protein A and G according to the immunoglobulin Isotype (14)

Species	Ig Isotype	Protein A	Protein G
Mouse	IgG1	+	+++
	IgG2a	+++	+++
	IgG2b	++	++
	IgG3	+	+
	IgM	Use anti mouse IgM	
Rat	IgG1	–	+
	IgG2a	–	+++
	IgG2b	–	++
	IgG2c	+	++
Rabbit	All isotypes	+++	++

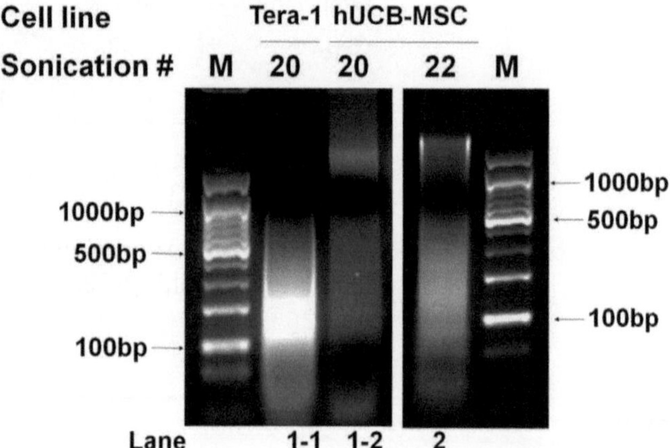

Fig. 1 DNA shearing in different cell lines and different cell number and volume. 1×10^8 cells in 5 ml SDS lysis buffer were used for each condition. 30-s-pulse sonication was applied to each sample as indicated number of sets with a Branson 450 Sonifier setup at output level 3. Tera-1, human testicular tumor cell line (*lane 1-1*) and hUCB-MSCs (*lane 1-2*) were sheared using same sonication condition, which is 20 sets of 30-s-pulse sonication. hUCB-MSCs of *lane 2* was sheared for 22 sets of 30-s-pulse sonication. "Output" means amplitude and "sonication #" means the number of sets of 30-s-pulse sonication. "M" means size marker, and corresponding sizes were indicated

sample and sonicate each sample containing 1×10^6 cells, respectively. If several binding proteins should be analyzed and used repeatedly in one sample, sonication condition of total cells in one sample can be set up in one tube to reduce the time for sonication (Fig. 1). Total cell numbers for ChIP analysis are calculated as follows:

$$[1 \times 10^6 \text{cells} / \text{sample}] \times [\text{sample number(target protein} + \text{negative control)}]$$

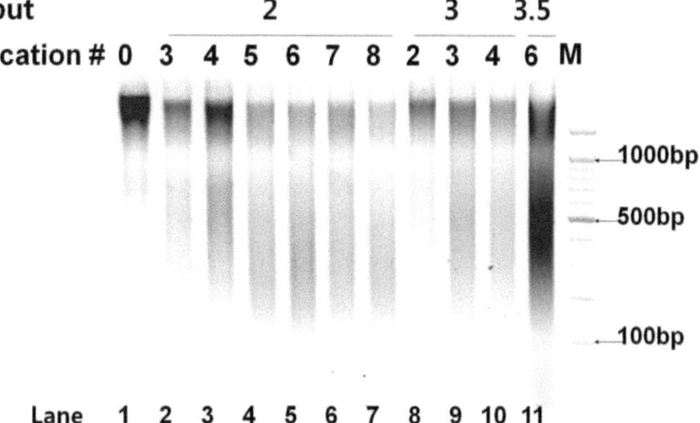

Fig. 2 Optimization of DNA shearing. Human umbilical cord blood derived mesenchymal stem cells (hUCB-MSCs) were used to setup sonication conditions. 1×10^6 cells in 200 μl SDS lysis buffer were used for each condition. "Output" means amplitude and "sonication #" means the number of sets of 10-s-pulse sonication. "M" means size marker, and corresponding sizes were indicated

Example If three different target proteins are to be analyzed, 4×10^6 cells would be needed for ChIP analysis. So, if you want to set up 4×10^6 cells condition, prepare $5 \times 4 \times 10^6$ cells for this.

6. Formaldehyde is very toxic when it is inhaled. Properly wrap the culture dish with foil in example before you put them into 37°C incubator. If you incubate them longer than recommended, unexpected binding between protein and DNA strand could be occurred resulting in false positive protein–DNA interaction.
7. Add protease inhibitors to PBS just prior to use. PMSF has a half-life of approximately 30 min in aqueous solutions. From step 5 in Subheading 3.1 to step 7 in Subheading 3.2, protein degradation should be protected. So, during these steps, all experiment should be done in ice-cold state and proteinase inhibitors should be added in incubation step.
8. Calculate SDS Lysis Buffer as 200 μl per 1×10^6 cells.
9. Vary the power setting and/or the number of 10-s pulses during sonication of the samples. Be sure to keep the sample on ice at all times (the sonication generates heat which will denature the DNA). Check the size of sonicated DNA by gel electrophoresis after reversion of cross-links. According to our experience, appropriated length DNA was obtained by shearing with 6–8 sets of 10 s pulse sonication using a Branson 450 Sonifier equipped with a 3 mm-microtip which was setup an output level to 2 (amplitude).
10. Assess fragmentation pattern and select the appropriate sonication condition. Fig. 2 shows the results of DNA shearing by

Table 2
Primer pairs for the p16INK4A promoter region

		Sequence	Size (bp)
p16INK4A—200 bp	F	ACCCCGATTCAATTTGGCAG	73
	R	AAAAAGAAATCCGCCCCCG	
p16INK4A—1 kb	F	CTCAAAGCGGATAATTCAAGAGC	66
	R	AAGCCTTAAGAACAGTGCCACAC	

various sonication conditions. As the output (amplitude) increase, the sonication efficiency also increase comparing Lane 2 and 9.

11. Enrichment of the ChIP target can be expressed as a fold difference between specific Ab-immunoprecipitated samples and those immunoprecipitated with no-Ab control.
12. Phenol:Chloroform:Isoamyl alcohol usually has an upper layer of buffer saturated water in the bottle. Do not use this buffer layer—you want the phenol/chloroform layer underneath.
13. Before recovering DNA, addition of an inert carrier, such as 20 μg glycogen or yeast tRNA, helps visualize the DNA pellet.
14. Promoters are adjacent to the transcriptional start site, where transcription of RNA begins for a particular gene. The core promoter includes DNA elements that can extend ~35 bp upstream and/or downstream of the transcription initiation site (13). Proximal promoter which contains primary regulatory elements is located approximately 250 bp upstream of the start site. Distal promoter which contains specific transcription factor binding sites is located further upstream. Primers for particular promoters are usually selected from transcription start site (TSS) to 1 kb upstream region. And the ideal product size of PCR is between 50 and 150 bp. Table 2 shows primer sets of the p16INK4A promoter.
15. In cases where the protein–DNA interaction takes place in a few cells (although the protein is widely expressed), more sample DNA will be required. Although the melting curve shows one peak in input analysis, the melting curve could not be good in enriched samples. In that case, increase the sample volume (2 μl, 4 μl, ...) for PCR analysis.
16. Although PCR band can be confirmed by agarose gel running, realtime-qPCR enables quantitative analysis of ChIP result.

References

1. Kang KS (2011) Epigenetic regulations in adult stem cells: the role of DNA methyltransferase in stem cell aging. Epigenomics 3:671–673
2. Li Z, Liu C, Xie Z et al (2011) Epigenetic dysregulation in mesenchymal stem cell aging and spontaneous differentiation. PLoS One 6:e20526
3. Chung SY, Hill WE, Doty P (1978) Characterization of the histone core complex. Proc Natl Acad Sci USA 75:1680–1684
4. Eickbush TH, Moudrianakis EN (1978) The histone core complex: an octamer assembled by two sets of protein-protein interactions. Biochemistry 17:4955–4964
5. Grunstein M (1997) Histone acetylation in chromatin structure and transcription. Nature 389:349–352
6. Ardehali MB, Mei A, Zobeck KL et al (2011) Drosophila Set1 is the major histone H3 lysine 4 trimethyltransferase with role in transcription. EMBO J 30:2817–2828
7. Koche RP, Smith ZD, Adli M et al (2011) Reprogramming factor expression initiates widespread targeted chromatin remodeling. Cell Stem Cell 8:96–105
8. Jung JW, Lee S, Seo MS et al (2010) Histone deacetylase controls adult stem cell aging by balancing the expression of polycomb genes and jumonji domain containing 3. Cell Mol Life Sci 67:1165–1176
9. Lee S, Jung JW, Park SB et al (2011) Histone deacetylase regulates high mobility group A2-targeting microRNAs in human cord blood-derived multipotent stem cell aging. Cell Mol Life Sci 68:325–336
10. Gilmour DS, Lis JT (1984) Detecting protein-DNA interactions in vivo: distribution of RNA polymerase on specific bacterial genes. Proc Natl Acad Sci USA 81:4275–4279
11. Costlow N, Lis JT (1984) High-resolution mapping of DNase I-hypersensitive sites of Drosophila heat shock genes in Drosophila melanogaster and Saccharomyces cerevisiae. Mol Cell Biol 4:1853–1863
12. Mukhopadhyay A, Deplancke B, Walhout AJ et al (2008) Chromatin immunoprecipitation (ChIP) coupled to detection by quantitative real-time PCR to study transcription factor binding to DNA in Caenorhabditis elegans. Nat Protoc 3:698–709
13. Smale ST, Kadonaga JT (2003) The RNA polymerase II core promoter. Annu Rev Biochem 72:449–479
14. Bonifacino JS, Dell'Angelica EC, Springer TA (2001) Immunoprecipitation. Curr Protoc Mol Biol Chapter 10:Unit 10 16

Chapter 13

Quantitative Fluorescence In Situ Hybridization on Paraffin Embedded Tissue

Mario Ricciardi, Mauro Krampera, and Marco Chilosi

Abstract

Quantitative fluorescence in situ hybridization (Q-FISH) is a complex technique for the quantitative evaluation of telomere length on cell preparations or on human tissues. The samples are stained with a fluorescent peptide nucleic acid (PNA) probe against the telomere oligonucleotides (sequence 5′-TTAGGG-3′). The measure of the telomere length is carried out using a fluorescence microscope equipped with a sensitive CCD camera and analyzing the pictures with a computer software that can perform fluorescence intensity measurements. Here, we describe the most used protocols to stain, acquire, and analyze fixed human cells in order to evaluate their telomere length.

Keywords Q-FISH, Telomere length, Fluorescence in situ hybridization, Peptide nucleic acid probes, Quantitative image analysis

1 Introduction

Human telomeres play critical roles, both in the maintenance of chromosomal stability, as well as in limiting the ultimate replicative capacity of cells. Telomere shortening has been suggested to be an important biological factor in aging, cell senescence, cell replication, cell immortality, and transformation to cancer (1). It is therefore important in pathology to have a method for quantitative assessment of telomere lengths applicable to individual human cells in archival material (2–5).

The invention of the Q-FISH telomere length has really improved the previous methods for the estimation of the telomere length classically based on Southern blot. In fact, the Q-FISH methodology provides estimate of telomere length in each individual chromosome with the resolution of 200 base pairs. In addition, Q-FISH may be used to estimate telomere length in species containing interstitial telomeric sites in their genomes and also has also been essential in estimating telomere length in the mouse, a species with ultralong telomeres difficult to measure using classical methods (5).

Kursad Turksen (ed.), *Stem Cells and Aging: Methods and Protocols*, Methods in Molecular Biology, vol. 976,
DOI 10.1007/978-1-62703-317-6_13, © Springer Science+Business Media, LLC 2013

The Q-FISH is based on the use of a fluorescent peptide nucleic acid (PNA) telomere oligonucleotide probe that generates stronger and more specific hybridization signals than standard DNA oligonucleotide probes (6, 7). In this context, the fluorescence intensity detected is directly proportional to the size of telomere repeats. This technique has been widely used and validated on metaphase nuclei and on whole nuclei sorted by flow cytometry (8, 9) and more recently also on frozen tissue, on paraffin embedded tissue and on fixed dropped cells (3–5, 10).

Here, we describe how to stain and analyze the telomere length with Q-FISH on paraffin embedded tissue, but the same technique can be used on frozen tissue and on fixed cultured cells, with some minor modifications.

When the Q-FISH is performed on paraffin embedded tissue the tissue blocks were sectioned at 4 μm. Briefly, deparaffinized slides are dehydrated through a graded ethanol series, placed in deionized water followed by deionized water plus 0.1% Tween-20 detergent. The slides are then steamed with citric buffer and allowed to cool at room temperature before the permeabilization treatment. At this point slides are treated with protease solution and then washed, dehydrated with alcohols solution and air-dried. After, slides are incubated with the probes, denaturated with heat and incubated to allow the hybridization of the probe. Once the hybridization as been carried out, the slides are counterstained with DAPI and mounted with prolong anti-fade mounting medium (2–5, 9, 10).

Later, fluorescence microscope equipped with a sensitive CCD camera and computer programs designed to control acquisition of digital images, as well as to perform fluorescence intensity measurements are required (2–5, 9, 10).

2 Materials

Prepare all solutions using ultrapure water (prepared by purifying deionized water to attain a sensitivity of 18 MΩ cm at 25°C) and analytical grade reagents. Prepare and store all reagents at room temperature (unless indicated otherwise). Diligently follow all waste disposal regulations when disposing waste materials.

- Slides with fixed tissue section.
- Incubation chamber adjusted at 75°C.
- Coplin jars.
- Xylene.
- Ethanol (100, 95, and 80%).
- Deionized Water.
- 10 mM sodium citrate, pH 6.5: 2.4 g/l sodium citrate tribasic, 0.35 g/l citric acid monohydrate.

- Phosphate-buffered saline (PBS), pH 7.2.
- TEN Buffer: 7.88 g Tris–HCl, 3.722 g EDTA, and 0.5844 g NaCl in 1 l of H_2O.
- Tween 20 buffer: 500 μl Tween 20/l PBS.
- 1% pepsin working solution (freshly prepared) in 0.01 M HCl.
- 10 mg/ml RNase solution: 100 mg RNase (Sigma) 10 ml sterile H_2O.
- Formamide Buffer: 70% formamide, 10 mmol/l Tris, pH 7.5.
- Cy3-labeled telomere-specific peptide nucleic acid (PNA): 0.3 g/ml PNA in Formamide Buffer.
- Water bath that can be adjusted up to 90°C.
- Incubator or water bath for incubation at 37°C.
- PNA wash solution [70% formamide, 10 mmol/l Tris, pH 7.5, 0.1% albumin).
- 4-6-diamidino-2-phenylindole (DAPI) (250 ng/ml in PBS).
- VECTASHIELD mounting medium.
- Confocal Microscope with DAPI/Cy3 filter set.

3 Methods

3.1 Deparaffinization of the Slides and Antigen Retrieval

1. Place slides at 75°C for 45 min in the incubation chamber;
2. Wash the slides two times, each time for 5 min in a Coplin jar containing fresh xylene at room temperature. Dispose of xylene in waste bottle in hood.
3. Wash slides by immersing successively for 3 min each in Coplin jars containing 100, 90, 85, and 70% ethanol at room temperature, with a final brief rinse in water to remove the traces of ethanol (see Note 1).
4. If tissue will not be stained immediately, air-dry and store frozen under nitrogen or in liquid nitrogen tank (storage in oxygen atmosphere will result in oxidation of DNA and reduced telomere staining).
5. Turn on the incubator and the water bath; set them respectively to 37°C and to 88°C.
6. Prepare 10 mM sodium citrate, pH 6.5, in a Coplin jar and prewarm in the water bath.
7. Warm 1% pepsin working solution to 37°C in a Coplin jar in incubator (see Note 2).
8. Incubate for 10 min the slides with tissue in 10 mM sodium citrate, pH 6.5, at 88°C (see Note 3). *This is the antigen-retrieval step.*

After this step the temperature of the water bath can be set to 78°C for next steps.

9. Rinse 1 min in a Coplin jar containing PBS, pH 7.2, at room temperature (see Note 4).
10. Briefly dip slides in Coplin jars containing 25%, 50%, and then 95% ethanol. Air-dry.
11. Treat for 2 min with 1% pepsin working solution in a Coplin jar in a 37°C water bath (see Note 5).
12. Rinse for 1 min with PBS, pH 7.2, at room temperature.
13. Briefly dip slides in 25%, 50%, and then 95% ethanol. Air-dry.

3.2 Telomere Staining

1. Add 80 μl of 10 mg/ml RNase solution to each slide, cover with coverslip, and let sit 10 min at 37°C in the incubator.
2. Remove coverslip and rinse slide 1 min in PBS (see Note 6).
3. Drain and then briefly dip slides successively in 25, 50, and 95% ethanol. Air-dry.
4. Add 25 μl of a Cy3-labeled telomere-specific peptide nucleic acid (PNA) and apply coverslip (see Note 7).
5. Heat slides 10 min at 78°C in the water bath that was preheated in Subheading 3.1. Cover or put slides in dark place to incubate overnight (at least 16 h) at room temperature (see Note 8).
6. Remove the coverslip from slide and place slide in a Coplin jar with room-temperature 70% formamide buffer to rinse for 15 min. Rinse four more times, 15 min each time, using fresh formamide buffer for each rinse.
7. Rinse four times in room-temperature Tween 20 buffer, 5 min each time.
8. Drain and then briefly dip slides successively in 25, 50, and 95% ethanol. Air-dry (see Note 9).
9. Add 80 μl DAPI working solution as a counterstain, apply coverslip, and incubate for 10 min at room temperature.
10. Rinse for 1 min in room-temperature PBS.
11. Briefly dip slides in 25%, 50%, and then 95% ethanol. Air-dry.
12. Add 10–20 μl VECTASHIELD, apply coverslip, and seal with clear nail polish to prevent drying out (see Note 10).
13. Store slides at −20°C until ready to take pictures using the confocal microscope (see Note 11).

3.3 Acquisition of the Image

1. Confocal Microscopy is required for the acquisition of the telomere staining. Fluorescence excitation/emission filters were as follows: Cy3 excitation, 546 nm/10 nm BP; emission, 578 nm LP; DAPI excitation, 330 nm; emission, 400 nm; (see Note 12).

2. Determinate the optimum exposure times, and held constant thereafter, such that all cells have an identical exposure time (see Note 13). This step must be also done at least on three control cell lines with a known telomere length in order to obtain a setup scale for the correlation of fluorescence and telomere length.
3. Determinate the extent of signal loss because of photo bleaching of the hybridized Cy3 telomeric probe by repeated acquisition of the same field (see Note 14).
4. Be sure to scan the slide in the *z-axis*, as well as the *x* and *y-axes*.
5. Take pictures and save images as TIFF files (for lossless compression) to enable Q-FISH analyses to be quantified using image-analysis software.

3.4 Analysis of the Image

1. Open the TIFF image with specific software for the quantification of the fluorescence.
2. Manually select the nuclei of interest (see Note 15).
3. Calculate the intensity of the fluorescence for the CY3 and DAPI channel.
4. For each nucleus, the individual Cy3 telomere signals were summed and this total value is divided by the total DAPI fluorescence signal for that nucleus (see Note 16).

4 Notes

1. Sometimes deionized water plus 0.1% Tween-20 detergent could be used in order to improve the permeabilization of the cells (2).
2. Some modifications to these protocols include the use of Proteinase K (10) or Protease Type VIII Solution dissolved in PBST at a final concentration of 0.5 mg/ml (2).
3. Antigen retrieval should be carried out also placing the jar with citric buffer pH = 6.5 in the microwave for a total time of 10 min at 100 W. In order to prevent the dry of the slides is recommended to stop and fill the jar with citric acid every 2–3 min. The refill should be done using a Citric Buffer Solution as warm as the one in use. For this reason is better to put the refill jar in the microwave too.

   Alternatively 0.2 M HCl for 25 min at room temperature.
4. Alternatively PBS+ 0.1% Tween 20 can be used to improve permeabilization (2); Coplin jar with slides and citric buffer should be left at room temperature for cooling down.

5. When different protease solution is used the time should vary between 1 and 5 min. Proteinase K is used 1:500 for 10 min at 37°C in TEN Buffer.
6. Be careful to remove gently only the coverslip; sometimes is recommended the use of a razorblade. A useful additional step should be added to reduce background fluorescence by incubating tissue sections 30 min in 70% formamide buffer with blocking reagent at room temperature. Before moving on, wash three times by briefly immersing the slide in PBS.
7. The amount of labeled telomere-specific peptide nucleic acid may vary; follow manufacturer's instructions. Amidate centromere probe (500 μg/ml): 5′-ATTCGTTGGAAACGGGA-3′ can also be used in conjunction with PNA probe.

   Cy-3 can be replaced with FITC or other available conjugated PNA.
8. Alternatively denaturation is carried out by incubation for 4 min at 83°C (3), or for 3 min at 80°C (10). Also the incubation time of the probe can change: from 2 h to overnight incubation.
9. Additional staining with other antibody should be performed at this point, before performing DAPI and anti bleaching mounting medium (2).
10. VECTASHIELD is very viscous; pipette slowly avoiding air bubble formation.
11. It is recommended not to leave the slides more than 1 week before capturing pictures.
12. If alternative dyes are used, excitation and emission frequencies may be adjusted appropriately. Images can be taken using a 40×, 63×, or 100× objective; 63× magnification objective lens with a numerical aperture of 1.4 and 100× magnification objective lens with a numerical aperture of 1.4 are recommended when improved resolution is necessary for histological assessment (2, 3).
13. Fluorescent microbead intensity standards should be used to confirm that telomeric signals were within the linear response range of the charge-coupled device camera. In order to reduce the analysis problems is very important to obtain a clear picture with no background (2).
14. The bleaching is calculated with software analysis and it's expressed in percentage of fluorescence loss per second.
15. A minimum of 10 nuclei per cell line/subline should be selected, especially for the control cell line (3).
16. This correction is done for potentially confounding differences in nuclear cutting planes and ploidy.

The type of tissue and the different preparation of the sample (cell, fixed tissue, etc.) should be considered before choosing the methods for the evaluation of the fluorescence. For tissue sections the most reliable method is the Background-corrected fluorescence offered by some specific software for Q-FISH analysis. This method compares the difference between the intensity of the brightest and the dimmest pixels in each nucleus: the dimmest 20% of red or green nuclear pixels are taken as representative of nonlabeled nuclear background, and the mean intensity of this background is subtracted from the average of the brightest 5%, 10%, or 20% of red or green pixels in that same nucleus. The resulting number is the corrected estimate of the average telomere intensity (3).

References

1. Blackburn EH (1991) Structure and function of telomeres. Nature 350:569–572
2. Meeker AK, Gage WR, Hicks JL, Simon I, Coffman JR, Platz EA, March GE, De Marzo AM (2002) Telomere length assessment in human archival tissues: combined telomere fluorescence in situ hybridization and immunostaining. Am J Pathol 160(4):1259–1268
3. O' Sullivan JN, Finley JC, Risques R, Shen WT, Gollahon KA, Rabinovitch PS (2005) Quantitative fluorescence in situ hybridization (QFISH) of telomere lengths in tissue and cells. Curr Protoc Cytom 12.6.1–12.6.15
4. O' Sullivan JN, Finley JC, Risques RA, Shen WT, Gollahon KA, Moskovitz AH, Gryaznov S, Harley CB, Rabinovitch PS (2004) Telomere length assessment in tissue sections by quantitative FISH: image analysis algorithms. Cytometry 58A:120–131
5. Slijepcevic P (2001) Telomere length measure ment by Q-FISH. Methods Cell Sci 23:17–22
6. Landsorp PM, Verwoerd NP, van de Rijke FM, Dragowska V, Little MT, Dirks RW, Raap RW, Tanke HJ (1996) Heterogeneity in telomere length of human chromosomes. Hum Mol Genet 5:685–691
7. Moyzis RK, Buckingham JM, Cram LS, Dani M, Deaven LL, Jones MD, Meyne J, Ratliff RL, Wu J-R (1988) A highly conserved repetitive DNA sequence, (TTAGGG)n present at the telomeres of human chromosomes. Proc Natl Acad Sci USA 85:6622–6626
8. Rufer N, Dragowska W, Thornbury G, Roosnek E, Lansdorp PM (1998) Telomere length dynamics in human lymphocyte subpopulations measured by flow cytometry. Nat Biotechnol 16:743–747
9. Poon SSS, Martens UM, Ward RK, Lansdorp PM (1999) Telomere length measurements using digital fluorescence microscopy. Cytometry 36:267–278
10. Ferlicot S, Youssef N, Feneux D, Delhommeau F, Paradis V, Bedossa P (2003) Measurement of telomere length on tissue sections using quantitative fluorescencein situ hybridization (Q-FISH). J Pathol 200:661–666

Index

A

B

C

Kursad Turksen (ed.), *Stem Cells and Aging: Methods and Protocols*, Methods in Molecular Biology, vol. 976, DOI 10.1007/978-1-62703-317-6, © Springer Science+Business Media, LLC 2013

N

O

P

Q

R

S

T

W

MIX
Papier aus verantwortungsvollen Quellen
Paper from responsible sources
FSC® C105338

If you have any concerns about our products, you can contact us on
ProductSafety@springernature.com

In case Publisher is established outside the EU, the EU authorized representative is:
Springer Nature Customer Service Center GmbH
Europaplatz 3, 69115 Heidelberg, Germany

Printed by Libri Plureos GmbH
in Hamburg, Germany